D0432370

There are Giants in the Sea

There are Giants in the Sea

Michael Bright

GUILD PUBLISHING LONDON

This edition published 1989 by Guild Publishing by arrangement with Robson Books

First published in Great Britain in 1989 by Robson Books Ltd, Bolsover House, 5-6 Clipstone Street, London W1P 7EB

CN 6705

Printed in Great Britain by St Edmundsbury Press Ltd, Bury St Edmunds, Suffolk.

Contents

Acknowledgements

For their help in the preparation of *There are Giants in the Sea*, I would like to thank, in alphabetical order, the following:

Sheila Anderson, Sea Mammal Research Unit, Cambridge, England; Martin Angel, Institute of Oceanography, Wormley, England; Chris von der Borch, Flinders University of South Australia; Bill Burton, *Baltimore Evening Sun*, USA; Miles Clark, *Yachting Monthly*, London, England; Malcolm Clarke, Marine Biological Association, Plymouth, England; Lucy Cacannas and Fergus Keeling, BBC Natural History Unit, Bristol, England; John Cloudsley-Thompson and Andrew Milner, Birkbeck College, London, England; Michael Frizzel, the Enigma Project, Maryland, USA; Richard Greenwell, University of Tuscon, Arizona, USA; David Heppell, Royal Museum of Scotland, Edinburgh, Scotland; Bernard Heuvelmans, Centre de Cryptozoologie, Le Bugue, France; Ray Ingle, British Museum (Natural History), London, England; Rosamond Kidman Cox, *BBC Wildlife Magazine*, Bristol, England; Paul Le Blonde, University of British Columbia, Canada; Brian Luckhurst, Division of Fisheries, Bermuda; Ray Manning and Clyde Roper, Smithsonian Institution, Washington DC, USA; Roy Mackal, University of Chicago, USA; Michael Malaghan, the *Skegness Standard*, England; C.E. O'Riordan, National Museum of Ireland, Dublin, Eire; Karl Shuker, West Bromwich, England; and Forrest Wood, US Naval Ocean Systems Center, California, USA.

For their eyewitness accounts, I would like to thank Owen Burnham, Northwood, Middlesex, England; Jack Bishop, Karen Flew, and Clyde and Carol Taylor, Maryland, USA; Robert and William Clark, San Francisco, California, USA; Dave Clarke, John Cock, Tony 'Doc' Shiels, and George Vinnicombe, Falmouth, Cornwall, England; C.D. Cole, Vancouver, Canada; John (Sean) Ingham, Bermuda; Captain Kingston Lewis, Clwyd, Wales; Marlene Martin, Marin County, California, USA; Alfred Peterson, Crayford, Kent, Englands; John Ridgeway, Scotland; Jim Thompson, Bellingham, Washington, USA; L. Welch, Weymouth, Dorset, England; and George Whiting, Larnaca Marina, Cyprus.

I am indebted to Bernard Heuvelmans for information about sightings recorded since 1966; that is, after his authoritive work on unknown sea creatures *In the Wake of Sea Serpents* had been published. I would also like to acknowledge the helpfulness of his comprehensive reference list.

1

The Mysterious Ocean

Are there gigantic unknown creatures living in the world's oceans, animals of which we have occasional sightings but which we cannot yet identify? Does the legendary leviathan, perhaps a giant sea serpent or a colossal octopus, lurk in the deep? There are certainly many monsters in the sea – giant whales and squid, enormous fish, jellyfish and shellfish, and sea turtles and seagoing crocodiles of great size.

The sea is a mysterious place. Almost three-quarters of the Earth's surface is covered by sea, 97 per cent of which is more than 200m (656ft) deep, and largely unexplored and unknown. The deep sea has been reluctant to reveal its secrets despite the best efforts of the few biologists and oceanographers who, down the centuries, have ventured into the depths simply to find out if anything is living there at all. Our knowledge is still minimal, and seemingly less even than the information gained from those barren, lifeless places, the Moon and Mars. An eminent marine biologist was once heard to remark that 'there are more footsteps on the Moon than there are on the sea-bed'.

Oceanographic research is difficult, because of the nature of the alien environment in which it must be carried out, and it is very expensive. Nevertheless, the sea is beginning to yield to the current invasion. A new generation of aquanauts, accompanied by side-scan sonars, underwater video cameras, deep-sea submersibles, bottom-hugging towing sledges, rectangular mid-water trawls that open and close at a chosen depth, drifting buoys, and orbiting, remote-sensing satellites in space, are beginning to build up a new picture of what is happening below the waves, particularly in the great depths.

At one time the deep sea was thought to be a dark, still and lifeless place. The pressures at the bottom of the ocean were considered to be too great for life to survive. Imagine the surprise, then, when a research vessel trawling on the bottom of the Philippines Trench, some 10,000m (33,000ft) below the ocean surface, brought up sea anemones, sea cucumbers and a bristle worm.

Today, we know that the deep sea is not perpetually calm. Just as there are weather patterns in the atmosphere, so too there are comparable systems under the sea. Deep-sea storms, containing far more energy than those at the surface, rage for weeks rather than days before subsiding. They scour the deep ocean floor, creating underwater blizzards, ripping up bottom-dwelling animals, and burying others in mountains of clogging, deep-sea mud.

Far from being devoid of life, the deep sea is full of wondrous animals, some carrying their own lights that shine brightly in the permanent night. There are the scavengers – rat-fish, brotolids and slimy hagfish – that feed on the debris raining down from productive surface waters. There are the predators – small viper-fish with glowing lures, and enormous six-gilled and sleeper sharks – that patrol the mid-waters and the ocean floor. In the very deep parts of the sea, the fishes are absent, leaving the small amount of relatively large food particles (about 3 per cent of that available at the surface) to the sea cucumbers, sea anemones and starfish.

And there have been even more surprising discoveries. In 1977, scientists were cruising an area to the north-east of the Galapagos Islands in the eastern Pacific, searching for underwater hot springs associated with cracks in the earth's crust, when they came across something quite unexpected.

Under the direction of Dr Robert Ballard, of the Woods Hole Oceanographic Institution, an unmanned underwater sled, with cameras and temperature sensors aboard, was put over the side of the research ship *Knorr* and towed back and forth over the target area. After six hours of watching and hoping, the instruments showed that the sled had passed over a hot water anomaly. It lasted for about three minutes. It was the only 'contact' in a search lasting 12 hours and over 16 kilometres of seabed. The underwater camera had taken 3,000 colour photographs.

The film was not removed for an agonizing couple of hours so that the camera, which had been in freezing temperatures at the bottom of the sea, could warm up, and prevent condensation from destroying the pictures. The first part of the film showed barren ocean floor covered by pillow lava, followed by younger ropy flows of lava. Another section of pillow lava marked the boundary line of low underwater volcanoes. Then, suddenly, for thirteen frames, the ocean floor was covered with hundreds of clams and mussels in a patch of misty-blue water. The thirteen frames coincided exactly with the temperature variation shown by the thermometers.

Later, the support ship *Lulu* brought the deep-sea submersible *Alvin* to the site. It took an hour and a half for the vessel to drift slowly and virtually blind to the ocean floor 2,700m (8,858ft) below. It arrived just a couple of hundred metres from the hot vent area and moved in slowly across the sea-bed.

First, *Alvin* passed over pillow larva, and the crew inside could see the same barren underwater seascape that had featured in the photographs but, as it approached the hot water, the researchers' attention was drawn to two purple sea anemones. Then, they noticed the water was shimmering, 'like air over a warm pavement'. As the submersible edged forward they saw the most incredible sight they had ever seen. Gushing from cracks in the lava was shimmering water that turned a dirty blue as minerals precipitated out and stained the rocks. All around were myriads of creatures, including giant clams and mussels up to 30 centimetres (12in) across that packed in dense clumps close to the vents, clusters of three metre-long red tube-worms, white crabs and squat lobsters, pink fish, the occasional purple octopus and shrimps with comb-like structures where their eyes should be.

Here was an amazing community of animals living on an oasis of rock just 50m (164ft) wide, on a hot-spring field in complete darkness. What, thought the researchers, could they be living on? The answer was quick to reveal itself. Samples taken back on board smelled of rotten eggs – hydrogen sulphide formed in the hot water gushing from the vents.

The sulphur provides food for bacteria, which, in turn, are consumed by some of the other creatures in the vent food-chain. Deep-sea communities usually depend on particles 'raining down' from above, but the hydrothermal vent ecosystem is unique, for unlike other communities on earth, it is independent of the energy provided by the sun. The energy comes directly from the centre of the earth.

In 1986 there was another intriguing vent discovery. The story started at the bottom of one of the other great oceans, the Atlantic. Another researcher from Woods Hole, Cindy Van Dover, found some blind shrimps at the bottom of the sea at a point in mid-Atlantic between Bermuda and the Azores. She noticed, though, that despite the lack of eyes they responded to light, and on closer examination, discovered that they had light-sensitive organs underneath their translucent shells. But what light, thought Van Dover, would the animal be reacting to in the great depths of the ocean where no sunlight can penetrate?

The answer came after a deep-sea expedition in the Pacific, led by Dr John Delanay of Washington University, came upon a strange glow about two kilometres (1¼ mi) down on the Juan de Fuca Ridge which lies 300 kilometres (186mi) to the west of Vancouver, Canada. There he discovered hydrothermal vents gushing very hot water that glowed in the abyssal darkness. Why it glows is not known, but no doubt these glowing vents also occur in the Atlantic providing a focal point for light-sensitive eyeless shrimps.

Many zoological discoveries of recent years have not involved enormous organisms but have nevertheless been significant and unexpected. Occasionally, we have to reassess the way we catalogue the animal kingdom. Some creatures just do not fit into the scheme that we have. The spherical, 10mm (½ in) diameter 'sea daisy', for instance, which looks much like a tiny upturned saucer, was found on submerged logs 1,000m (3,300ft) down in the sea off New Zealand in 1985. It was assigned not only to a new species, but also to a new class of its own.

It has been suggested that about three-quarters of all the different types of fish that swim in the sea have yet to be discovered. During exploration dives amongst the coral reefs of the Bahamas by scuba divers during the early 1970s, about 40 per cent of the fish they found were new to science. Beyond the shallows of the continental shelves, out of the reach of commercial trawls and drift nets, there are likely to be many more living in the ocean depths. The small nets, traps and trawls of research vessels are easily avoided by even the slowest swimmers.

Imagine a visitor from another planet cruising six kilometres above the Earth and reaching down for samples of terrestrial life-forms with a large butterfly net on a long piece of string. Is it likely that these explorers, fishing for a representative sample of life on Earth, would net anything of significance? Maybe they would catch the occasional bird or insect, and perhaps scoop up a few slow-moving mammals, a handful of bushes and a few tufts of grass. With such an ineffective way of sampling they would never come anywhere near to discovering the diversity of living things on this planet, let alone have any understanding of the relationship between animals, plants and the places in which they live. Yet, that is precisely how we have been looking for creatures, both large and small, in the world's oceans.

It is, perhaps, understandable that tiny planktonic organisms or small deep-sea fishes will escape the gaze of man and go undetected and undescribed by science, but it is sobering to find that even the largest marine creatures are travelling the oceans of the world mostly unseen and unheard by the human observer.

Some species of beaked whales, for example, are described from just three or four carcasses that have been washed up on remote shores. Andrews' beaked whale *Mesoplodon bowdoini* is known only from a few specimens that have been stranded on New Zealand beaches. Hector's beaked whale *Mesoplodon hectori* is recognized from strandings in New Zealand, Tasmania, California, the Falkland Islands and South Africa, Longman's beaked whale *Mesoplodon pacificus* is only known from three skulls – two found on Australian beaches and one which was spotted in Somalia.

Yet, these are large marine mammals. Some are many metres long and potentially conspicuous, for they must come up to the surface regularly in order to breathe. Nobody has spotted their snout, back or blowhole during that brief but necessary moment at the surface. Nobody has ever seen them alive. Their distribution, biology and behaviour is totally unknown. If they are so rare, one wonders, how does one beaked whale find another in the vastness of the ocean? Are there special beaked whale rendezvous sites which we have yet to discover? Where are the young born and brought up, safe from the prying eyes and open jaws of the ocean's predators?

And there are many more elusive cetaceans (whales and dolphins) which the sea has presented to us dead on a beach or allowed us an occasional sighting. The 2m (6.5 ft) long spectacled porpoise *Phocoena dioptrica*, is thought to live in the sea between the east coast of South America and the Falkland Islands but it has never been seen alive. The Atlantic humpback dolphin *Sousa teuszii*, from the tropical waters of West Africa, is known from just six individuals. Until 1970, Frazer's dolphin *Lagenodelphis hosei* was only known from one carcass washed up on a Malaysian beach in 1895. And the pygmy right whale *Caperea marginata*, weighing in at five tonnes and measuring six metres, is described from several stranded specimens and the occasional individual harpooned from whaling ships. One mysterious cetacean, which has a tall, erect dorsal fin like a killer whale but is clearly not one, is only known from four sketches drawn by the Antarctic explorer and naturalist Edward Wilson in 1902.

The Belgian zoologist, Bernard Heuvelmans, in a paper in *Cryptozoology*, collected together a checklist of unidentified creatures thought to be living in the world's oceans and came up with no fewer than eight species of whales and dolphins for which we have solid but scanty knowledge. He cites a high-finned sperm whale, 18.3m (60ft) in length, that was often seen off the Shetland Islands in the seventeenth century. No less than the 'father of cetology' (the study of whales and dolphins), Sir Robert Sibbald spotted one in 1692.

There was also a school of 9.1m (30ft) long, black-and-white beaked whales which the nineteenth-century naturalist, Philip Gosse, watched for several hours in the North Atlantic. At some time during the first part of the eighteenth century, a twin-dorsal-finned, black-and-white spotted dolphin was seen by Antonio Mongitore in the Mediterranean off the Sicilian coast. The anterior dorsal fin looked like a horn set on the top of the head. The scientific community greeted the observation with disbelief, but in October 1819 a school of these strange dolphins appeared before an expedition led by Jean Quoy and Joseph Gaimard which was sailing between the Sandwich Islands and New South Wales. And, more recently, in his book *Field Guide of Whales and Dolphins* published in 1971, Captain Willem Morzer Bruyns describes what he calls his 'Alula whale', a brownish-coloured killer whale with white star-shaped scars that frequents the eastern part of the Gulf of Aden, north of the village of Alula. It is 6.1m (20ft) long with a 0.6m (2ft) tall dorsal fin.

Unidentified whale spotted in the Antarctic in 1902. *From sketches by Edward Wilson.*

All these whales and dolphins have been seen by reliable, informed witnesses, but without more sightings and information, they cannot be officially recognized by the taxonomists.

Of the 78 known species of whales and dolphins, four have been discovered and recognized by science within the last hundred years. (Indeed, between 1900 and 1983, 134 species of mammals, found on land and sea, have been newly discovered. The largest terrestrial mammal to have been discovered in recent years was the kouprey, an ox-like animal that lives in Indo-China. It was recognized in 1937. One land mammal, Bulmer's fruit bat *Aproteles*, had the doubtful distinction of being recognized as a fossil and thought to be extinct, rediscovered in a cave in New Guinea, and exterminated by hunters, all during the same year!)

In the world's oceans, large animals sometimes turn up that are unexpected. In recent years, two such surprises confronted science.

On 15 November 1976, the US research vessel *AFB-14* was about to get under way from its station off one of the Hawaiian islands, and was hauling in two enormous parachute-like sea anchors from a depth of 165m (540ft), when the crew realized they had accidentally caught some large creature of the deep. Trapped inside one of the 'chutes was a very large shark, about 4.46m (14ft 7in) long and weighing 750kg (1,650lb). Its great blubbery lips, surrounding a broad gape set on protruding jaws, instantly gained it the nickname 'megamouth'.

The strange fish was hauled on board and taken ashore. Scientists who examined it gave it the scientific name *Megachasma pelagios* – 'yawning mouth of the open sea'. It is thought to be a slow-swimming filter feeder, but unlike the whale shark and basking shark which skim surface waters for plankton, megamouth probably swims, with jaws agape, through patches of deep-sea shrimps. There is speculation that the shrimps are enticed towards the maw with the aid of bioluminescent spots around the mouth.

In November 1984, another megamouth was netted by commercial fishermen off Catalina Island, near Los Angeles. Fortunately there was a fisheries officer aboard at the time and he recognized it as something special. The shark was taken to the Los Angeles County Museum. Interestingly, it, like the Hawaiian specimen, is a male. Female sharks generally grow larger than males, so might there be even larger megamouths living in the depths of the Pacific?

One of the scientists who was involved with research on megamouth, Leighton Taylor, of the Waikiki Aquarium, was reported as saying: 'The discovery of megamouth does one thing: it reaffirms science's suspicion that there are still all kinds of things . . . very large things . . . living in our oceans that we still don't know about. And that's very exciting.'

There had never been any hint that megamouth exists: no mariners' tales or native folklore, no surprise encounters, and no tantalizing glimpses from underwater cameras. However, another strange fish – a bony one – was regularly caught by fishermen in the Comoro Islands, yet was thought by fish biologists to have become extinct, along with the rest of its type, about 70 million years ago. It is the coelocanth.

The distinctive bluish-grey fish with white patches, about 1.5m (4ft) long, was hauled up in nets from a depth of 70m (230ft), about five kilometres from Chalumna Point at the mouth of the Chalumna River, south-west of East London, in South Africa in 1938, by Hendric (Harry) Goosen of the deep-sea trawler *Algoa Bay* (the *Nerrine* in some accounts). He saw that it was unusual and did all in his power to preserve it and get it back to land. The trawler carried an aquarium tank but the coelocanth was too big – 1.5m (5ft) long and weighed 57.6kg (127lb). Goosen had to pack the fish in ice. He then radioed base – the offices of Messrs Irvin and Johnson – and asked them to alert the local museum. Miss Courtenay Latimer, the first full-time curator of the East London Museum, had encouraged local commercial and sports fishermen to look out for unusual specimens with which she could build up the museum's fish collection. She was in the habit of visiting the fish market when the boats returned and got to know the skippers and crews. She arrived at the cold store to find a pile of sharks and a large scaly blue fish. The fishermen had not seen anything like it before.

Eventually she took the fish back to the museum in a taxi. The Chairman of the Board of Trustees was not very impressed, declaring the animal to be a 'a freak and nothing else'. But, nevertheless, he humoured Miss Latimer, who by this time was convinced that she had something unusual and perhaps important, and agreed that the fish could be kept. It was first preserved in formalin (it was beginning to get a little 'high') and later mounted, albeit rather badly, by a local taxidermist.

Miss Latimer sifted through her books and concluded that the fish was like 'a lung fish gone barmy'. Fired with renewed enthusiasm and intrigued that she could not find the fish in any textbooks she made some sketches and sent them to the distinguished ichthyologist (fish scientist) Professor J L B Smith of Rhodes University in Grahamstown. He did not reply immediately and so the rotting internal organs, which had not taken up the formalin, were thrown away. Eventually, he sent a telegram suggesting that the skeleton and gills should be kept. When he reached East London, he immediately recognized the fish (although upset at the way it had been treated) as something special. 'I always knew,' he said at the time, 'somewhere, or somehow, a primitive fish of this nature would appear.'

The fish resembled, almost exactly, fishes that were swimming the seas between 350 and 70 million years ago. It was a living fossil.

Goosen, who had a permanent scar on his hand where the fish had bitten him, was once heard to remark: 'I'm the only man that was ever savaged by a fossil.' Smith was quoted as saying: 'I would hardly have been more surprised if I had met a dinosaur in the street.'

The scientific community, meanwhile, accepted the creature's new scientific name *Latimeria chalumnae*, but to Smith it became known as 'Old Fourlegs'. For Ogden Nash it meant the source of another rhyme:

> It jeers at fish unfossilized
> At intellectual snobs elite;
> Old Coelocanth, so unrevised,
> It doesn't know it's obsolete.

The first specimen had been caught on 22 December 1938. The second, caught fourteen years later, almost to the day – 24 December 1952 – was hauled up on a fishing line from 200m (656ft) down by a local fisherman off Anjouan, one of the Comoro Islands, in the Mozambique Channel. After the first find, Smith had searched the eastern seaboard of Africa for more specimens, distributing pamphlets describing the fish, putting up posters, and offering a reward of £100. He was understandably excited when the reports of a second fish came in. He asked the South African government for help in reaching the Comoros quickly and the then Prime Minister, Dr Malan, personally ordered the South African Air Force to help Smith recover the fish and bring it back to the University. The islands were French possessions at the time and the authorities were so miffed that Smith had flown in and taken the fish without as much as a by-your-leave that he was banned from visiting the Comoros again.

Since then about 135 specimens have been caught, the next 83 going to French institutions. Consequently, most anatomical work has been by French scientists. Non-French researchers have had severe restrictions on what they can do with specimens; in some cases a $5,000 price-tag has been put on simply the loan of a specimen, and scientists have had to agree not to publish data until French scientists have done so first.

The Comoro islanders, though, have been unaware of the treasure at the bottom of their sea. Indeed, they have not been at all impressed with Old Fourlegs. They had been catching coelocanths, known locally as *Kombessah* or *Mame*, for decades; two or three, on average, a year (although Japanese fishing consultants have updated local fishing methods with the result that 10 – 15 have been caught in recent years).

The catching of coelocanths is incidental to oilfish, *Ruvettus pretiosus*, fishing. Small outrigger canoes go out at night and handlines are put out. The line itself is tied to a rock and sunk down to the bottom at 200 – 300m (656 – 984ft). The line is jerked and the stone released, leaving the line, with its hooks baited with pieces of 'rudi', *Promethichthys prometheus*, trailing freely in the depths.

Once hooked, apparently, the fish is an infernal nuisance. It is very powerful and difficult to bring up to a small boat, such as a canoe. It has a powerful bite and snaps repeatedly at any limbs within reach of its mouth. And, even if it reaches the surface, and is dispatched with a blow to the head, it does not make good eating. Some people eat the oily flesh if it is lightly salted. The rough scaly skin, however, is very useful. It is used as a kind of 'sandpaper' for roughing-up the inner tube when mending a puncture on a bicycle.

The discovery of living coelocanths showed that this particular line has changed very little for several million years, but until 1955 nobody had seen the fish alive. The learned journal *Nature* carried a report on 26 February by Jean Millot. He had tried to catch a coelocanth for observation:

> Throughout the night – which the delighted population of Mutsamadu passed singing and dancing to celebrate the capture – the Coelocanth was watched over with admirable care. It seemed, although quite bewildered at the sequel to its ascent to the surface, to be taking the situation very well, swimming slowly by curious rotating movements of its pectoral fins, while the second dorsal and anal, likewise very mobile, served with the tail as a rudder.

And that was all they were able to see, for the fish died the following afternoon. Nevertheless, the few observations that Millot was able to make did serve to tantalize marine biologists, but still nobody had seen the creature living in its normal environment 200 – 300m (656 – 984ft) down at the bottom of the Indian Ocean. That is until early in 1987, when films and videos, taken from a deep-sea submersible by a German team of scientists headed by Professor Hans Fricke, of the Max-Planck-Institut für Verhaltensphysiologie at Seewiesen, revealed something of the secret life of the coelocanth. The observations have completely changed our view of the fish and, as is often the case in science, thrown up a whole new set of intriguing questions.

The submersible *Geo* performed 40 dives at 30 different locations around the entire coast of Grande Comore and the north coast of Anjouan. On 17 January at nine o'clock in the evening the team found a coelocanth and they became the first people to see the creature alive in its natural environment. A total of six coelocanths, estimated to be 120 – 180cm (47 – 71in) long, were observed and filmed off a two-kilometre stretch of the Grande Comore coast. Two of the fish were sitting on the bottom and the other four were moving about slowly.

Before Fricke had made these recent observations, it was thought that the lobed paired pectoral and pelvic fins were used to crawl across the sea-bed, in the manner of some early amphibian. This is now known not to be the case. Indeed the fish does not even use them as an 'undercarriage'; it simply flops down on its belly, folding the fins against the body. To swim, it sculls with the paired fins like a swimmer using his arms in a 'crawl' stroke, except that the fins move more like a lizard or horse with the front-left fin moving together with the back-right and vice versa. The fish also makes use of upwellings, using the fins as hydrofoils to stabilize itself in the water. So, although walking is not observed in the fish itself, the action could have facilitated the transition, several million years ago, to locomotion on land. The coelocanth may not be the missing link between fish and land vertebrates, but it certainly shows some of the right features.

There was, however, some curious behaviour yet to be explained. All the fish observed by Fricke would perform a headstand and stay in that position for about two minutes. Two fish were even seen resting upside-down with the belly facing the surface. Others swam backwards.

The headstand behaviour can be explained, perhaps, as a way of detecting prey fish under the sand on the sea floor using electroreception. Sharks have a similar system. Electroreceptors in the shark's snout, known as the ampullae (pits) of Lorenzini, are able to detect the minute electric currents produced by contracting muscles in the prey animal, and thus indicate where a camouflaged plaice or flounder is hiding under the sand. The coelocanth is known to have a rostral organ in the head which might work in the same way and so, by performing the headstand, the fish could be directing its electroreceptor at the sea-bed in search of food. Unfortunately, the researchers were unable to witness any feeding. They were, however, able to generate prey-like electrical signals from some electrodes on the hull of the submersible and, in true Pied Piper fashion, encourage the coelocanths to follow the sub.

Fricke believes that the Indian Ocean is not the only site for coelocanths, and that they may exist elsewhere, in the Atlantic off Madeira for instance. It is likely that the Comoro Islanders came across the coelocanth because they are one of the few fishing communities to put their lines down so deep. If fishermen in other areas did the same, then maybe they too would come to market with a 'living fossil'. Fricke's research continues.

In the meantime there is concern for the continued survival of the coelocanth. The species has lived with little change for millions of years, but today it is this very fact that has made it valuable both to researchers and to exhibitors. People now want to capture a coelocanth alive and put it into an aquarium that can be set at the same pressure and lighting conditions to which the fish is exposed in its natural habitat. Unfortunately, this is likely to put a premium on coelocanth captures, and once one aquarium has a living coelocanth others will follow. Very quickly, the population would be wiped out for it is thought that the reproductive rate is slow and the population localized.

Some of those hooked in the past have been pregnant females containing eggs. But in 1975, a specimen was dissected at the American Museum of Natural History by Dr C Lavett Smith together with Professor C S Rand, of Long Island University, that was to prove to be an important breakthrough. The breakthrough though, had had to wait for some time because the US specimen was subject to the French restrictions – what one of the researchers described as 'a gentleman's agreement not to publish before the French had finished their anatomical work'.

The fish had arrived in New York in 1962 and was put on display. Nobody began to do serious work on it until thirteen years later; and that work proved to be some of the most important to date. The specimen that had been sitting on the shelf contained five well-developed embryos. This confirmed what researchers had suspected. The coelocanth does not release large numbers of eggs like other bony fish, but gives birth to live young. This could also go some way in explaining its success in surviving all this time. This reproductive strategy, however, would not be conducive to heavy exploitation by man. It is likely we would wipe it out in no time. Professor Michael Bruton, of the J L B Smith Institute, who led many expeditions to search for the coelocanth, is concerned that we could be responsible for its downfall:

> It would be ironic if, after all the fuss made about the coelocanth and its value historically, man caused it to become extinct.
>
> There is a real threat that the coelocanth may become vulnerable. It has a restricted distribution, a low fecundity, and there is a big reward on its head.

Let us hope that sense will prevail and that the coelocanth is protected. There is in existence a society bent on preserving the species, and the Commission for International Trade in Endangered Species (CITES) has categorized it with animals that cannot be traded commercially, but only for scientific research. Its future, therefore, is in the hands of the very people who can ensure its safety. It will be interesting to see if they measure up to the task.

So, if our knowledge and understanding of animals already being studied, like sharks, whales and coelocanths, is so fragmentary, and if large creatures continue to appear that are totally new to science, what of the other unknowns, the legendary monsters?

Throughout the history of our exploration of the sea, stories of strange animals and tales of unexpected encounters have been commonplace. The Old Testament of the Bible has several references to the dragon of the sea. The folklore of the ancient Norsemen, the Aborigines of Australia, the Inuit Eskimos and the North American Indians is full of sea serpent stories. Some, like the story of the kraken, have been taken out of mythology and placed firmly in the scientific textbooks. Might, then, the legends associated with the giant sea serpent be transformed, by scientific observation and scrutiny, into reality? One piece of evidence is exciting, if not a little frightening.

Squid-like sea monster on Scandinavian rune stone.

In 1950, Anton Bruun headed the famous *Galathea* oceanographic expedition from Denmark to explore the deep sea. During the build-up to the departure Bruun had surprised the general public and scientific community alike by declaring that one of the animals on his shopping list was a great sea serpent. Then, to an incredulous audience, he revealed a piece of information that he and his mentor

Johannes Schmidt, who had discovered that common eels spawn in the Sargasso Sea, had kept to themselves since 1930.

In that year the two scientists were aboard the *Dana* in the southern part of the Atlantic Ocean – latitude 35° 40′ south, longitude 18° 37′ east to be precise. This put them on the western edge of the Agulhas Plateau, off the Cape of Good Hope at the southern tip of Africa. They were trawling at depth for new marine specimens but were in for a shock when they hauled up one net from 150 fathoms. In it was a specimen that could give some credence to serpents' tales – an eel larva.

The eel starts out in life as a small, thin leaf-like creature known as a leptocephalus, and it is usually about a tenth the size of the adult. A common eel *Anguilla anguilla,* for example, has a leptocephalus larva just seven centimetres (3in) long. Imagine then the surprise of Schmidt and Bruun when they hauled out of the ocean a leptocephalus, not centimetres long, but nearly two metres! It has 450 vertebrae – three times as many as the largest known eel. This suggests an adult eel with the staggering length of 30m (98ft).

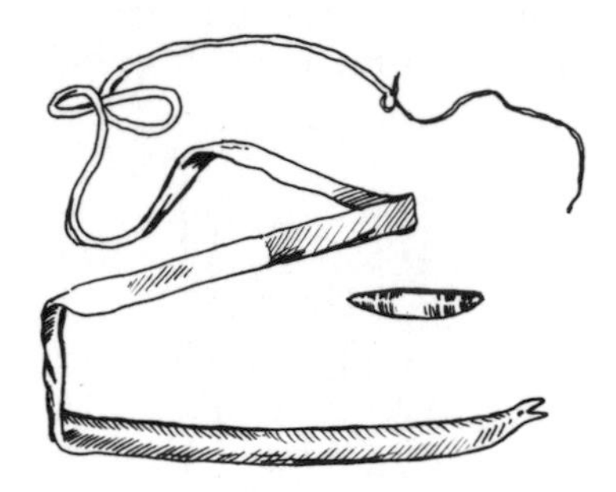

The giant eel larva found on the *Dana* expedition in 1930 compared to a common eel larva.*From a photograph.*

It is, perhaps, a little fanciful to propose that the ratio between the length of larva and the adult follows the same ratio as with other eels. In fact, the *Dana* larva resembles other larvae, such as *Leptocephalus giganteus,* caught in surface waters off south-west New Zealand, in the Straits of Florida, and off South Africa. They are thought to be the larvae of notocanth fishes – deep-sea spiny eels that are not eels at all but are eel-like in shape, and only grow to two metres long at the most. In the mid 1960s, a *L. giganteus* specimen was caught in the process of metamorphosing into the adult stage and it could be identified clearly as a notocanth. On the other hand, the ichthyologists could have got it all wrong. In some of the reference works I have examined, creatures labelled snipe eels (family Nemichthyidae) look remarkably like the *Dana* larva (one marine biologist proposed this to be the case). So, if this identification is in doubt, why could we not also consider the more exotic and exciting explanations? Bruun, a leading marine biologist, was prepared to speculate, why not others?

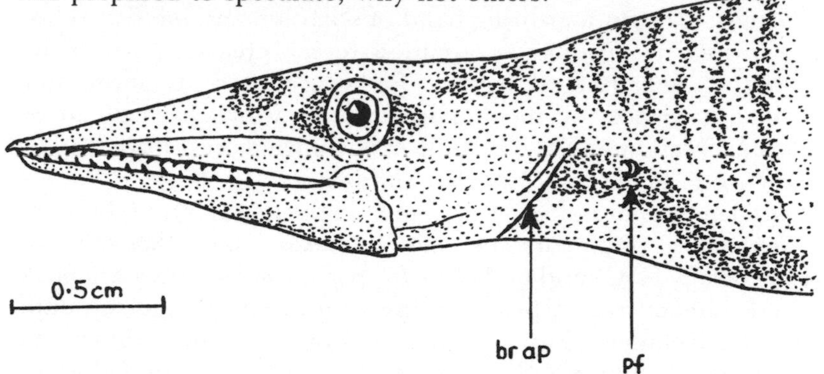

HEAD OF HOLOTYPE of *Leptocephalus giganteus*
br. ap: GILL OPENING pf: PECTORAL FIN

There is no doubt that animals can grow to a tremendous size in the sea, their bulk supported by the water. The largest creature ever to have lived on the planet – the blue whale – swims in the sea to this day, albeit only just, because of overfishing on the part of the commercial whalers during the first half of this century. It is several times bigger than the largest known dinosaur and many times bigger than the largest elephant or mammoth. Nevertheless, its life is very little understood; its breeding grounds are unknown; the routes it takes during migration unclear; and how well its populations are recovering after the great whale slaughter is difficult to assess. Even though it is an immense animal, the blue whale lies hidden, for most of its life, below the surface.

Other marine animals and plants also grow to huge sizes: giant bony fish such as the sturgeon, the marlin and the sunfish; giant cartilaginous fish such as the whale shark (the largest fish in the sea), the great white shark (the largest predatory fish), and the manta ray; giant reptiles such as the leatherback turtle, the Komodo dragon, and the estuarine crocodile; cephalopods such as the Pacific octopus and the giant squid; and giant crustaceans such as the Japanese spider crab and the Alaskan king crab. They all have one thing in common – despite their enormous size, we know very little about them.

So, if there are so many giants living in the sea, why should there not be a giant sea serpent? Why are many scientists reluctant even to consider the subject at all? Some researchers believe that if the giant sea serpent did not exist, we would have to invent it. We have, perhaps, a deep psychological need for sea monsters to be living in the darker, unknown parts of the oceans, much as we 'need' ghosts, UFOs, and little green men from Mars.

There is, though, a growing band of scientists and interested lay-people which is turning its attention to the creatures which are featured in folklore and legend, for which there is substantial anecdotal evidence for their existence, but which are still to be ratified as 'real' animals. The subject is known as cryptozoology – the science of hidden or unknown animals.

The term 'cryptozoology' was coined in the late 1950s by Bernard Heuvelmans, president of the International Society of Cryptozoology (ISC) and author of *In the Wake of Sea Serpents,* a book which documents sea serpent sightings from ancient times up until 1966. Heuvelmans has become known as the 'father of cryptozoology', but the honour of the title of 'grandfather of cryptozoology' he acknowledges should go to the Dutch zoologist Antoon Cornelius Oudemans who published the famous monograph *The Great Sea Serpent* in 1892. Heuvelmans feels that this was the starting-point of the science of cryptozoology. Oudemans had put the subject on the scientific map much as Chladni brought respect to studies of meteorites and Cuvier started the science of palaeontology.

The society itself was set up in January 1982, and is based in Tucson, Arizona. 'The ISC serves as a focal point,' according to its prospectus, 'for the investigation, analysis, publication, and discussion of all matters related to animals of unexpected form or size, or unexpected occurrence in time or space.' It has been the focal point for debate on the Yeti and Sasquatch (wildmen from Asia and North America), Mokele-Mbembe (a living dinosaur in the Congo), and Nessie and Champ – freshwater lake monsters from Scotland and New York State respectively. The Society also considers those animals, such as the Tasmanian tiger, that are thought to be extinct but which could be alive and well and living in remote areas.

The society's members have had some success. The *onza,* a long-legged puma-like cat has been found and is being studied in Mexico. The mysterious mermaid-like *Ri* was firmly established as the dugong.

Interestingly, the ISC is slowly representing the acceptable face of the hitherto slightly quirky study of unknown animals. Many respectable scientists have become members, among them eminent oceanographers and marine biologists who are tackling the thorny questions about sea serpents and other sea monsters.

Professor Paul LeBlonde, an oceanographer from the University of British Columbia, for example, commissioned a study of reports of unusual sea creatures spotted off the Pacific coast of Canada. Forrest Wood, of the United States Navy's Ocean Systems Center in San Diego, along with Dr Joseph Gennaro, of New York University, and Professor Roy Mackal, of the University of Chicago, investigated the case of the giant octopus *Octopus giganteus*. Professor Roy Wagner, of the University of Virginia, drew attention to the *Ri,* a mysterious aquatic animal from Papua New Guinea, and Thomas Williams and his colleagues from the Ecosophical Research Association solved the puzzle. Dr Eric Buffetaut, of the University of Paris, considered the vertical flexing of ancient crocodile-like creatures as a way of explaining the strange form of locomotion mentioned many times by sea serpent eyewitnesses. Dr George Zug, of the Smithsonian Institution in Washington DC, was one of the many distinguished scientists who reviewed, among other things, a videotape of the monster of Chesapeake Bay.

In the past, many eminent scientists and explorers have supported the existence of gigantic unknown creatures in the sea. Sir Joseph Banks (1744 – 1820), who sailed the world with Captain Cook and who was a president of the Royal Society, had 'full faith in the existence of our Serpent in the Sea'. Thomas Huxley (1825 – 95), a leading naturalist of his time who firmly backed Darwin's *Origin of Species* and clashed with the Bishop of Oxford, Samuel Wilberforce, in the famous British Association debate and who was another President of the Royal Society, felt that there was no reason why enormous snake-like sea creatures should not exist.

Oudemans, an authority on ticks and mites and a member of a very distinguished scientific family, wrote, 'I see no reason why dogma and prejudice should blind scientists to the possible existence of the Great Sea Serpent. The great wastes of water which cover the surface of our planet are largely unsounded and unexplored. Who is to say that somewhere in those mysterious deeps there does not lurk the unknown animal which so many witnesses have claimed to see?'

And more recently, Anton Bruun put cryptozoological matters firmly on the map when, at the XIV International Congress of Zoology in Copenhagen in 1953, he described a series of creatures living at great depths, including the giant eel larva, and then asked, to the amazement of his fellow zoologists, 'If a chordate [animals with a stiff "backbone" including the vertebrates with a true backbone] can live at the bottom of the sea, why not a sea serpent?'

The study of sea serpents and other unidentified 'monstrous' marine creatures is clearly not something that should be assigned simply to the eccentrics and cranks in the scientific community.

Today, these are legitimate, although not mainstream, studies.

In this book, I aim to reflect the renaissance in sea monster research. First, I describe what we might consider 'the mysteries' – the recent and not so recent sea monster sightings from around British shores, on the east and west coasts of North America, and from elsewhere in the world. Second, I describe the controversial evidence for the existence of the giant octopus and present the incontrovertible evidence for the giant kraken. And third, I review some of the explanations for the mysterious sightings, look at some of the real giants of the sea which might be mistaken for something far more exotic, and consider some of the ancient creatures from the fossil record which, like the coelocanth, might turn up once more on our doorstep.

For the privileged few who have glimpsed, often for no more than a few seconds, strange and gigantic creatures swimming in the sea, the sea serpent is a real phenomenon. 'Sea serpents,' they say, 'are alive and well.' There had been, however, some doubts.

There had been a feeling among cryptozoologists that sea serpent sightings had dropped off in recent years, the animals frightened away by noisy bulk carriers and supertankers. But this is not the case. Bernard Heuvelmans, relying on information he has gleaned from newspapers, magazines and personal correspondence, reveals that the number of reported cases has remained relatively steady for the past 150 years. In the first few chapters we look at just a few of the more exciting encounters.

2
The British Experience

Surrounded as it is by the sea, it comes as no surprise to discover that the British Isles have had their fair share of sea monster reports. Indeed, there was a glut of sightings and great debates in Victorian times. It was fashionable among the middle classes to go 'monster spotting'. One commentator – Ronald Binns in his book *The Loch Ness Mystery Solved* – even considered that the enigmatic Miss Woodruff in John Fowles' *The French Lieutenant's Woman* was really spotting sea serpents from the end of the Cobb at Lyme Regis. She invented the Frenchman to disguise the fact from her admirer Mr Smithson, a follower of Darwin.

Bernard Heuvelmans, examining the sea serpent sightings from newspapers and magazines during the nineteenth century, found that two-thirds of those reported from 1850 to 1890 were from British sources. Many were from ships visiting the farthest corners of the Empire, but some were on home shores. And the sightings continue to hit newspaper headlines to this day. Taking you on an anti-clockwise journey around the British coast, I have gathered as many of the eyewitness accounts, both old and new, that I can find. We start in the south-west peninsula of England – in Cornwall.

Cornwall has traditionally been the place of shipwrecks, smugglers and sea monsters. It is a place where myths and legends abound about the sea and the folk who live off it. There is the *hooper* of Sennen Cove, a spirit which appears in the form of a bank of fog. There are the sailors on the beach at night who do not reply when they are spoken to. They have wrinkled hands and squelching boots and they are the souls of shipwrecked sailors seeking their ancestors. Then, there was the giant Wrath who lived in Ralph's cupboard near Portreath. He watched for shipwrecks and devoured the unfortunate crews.

In recent years, a sea monster has dominated quayside conversations along the south Cornish coast. It is known by its Cornish name of 'Morgawr', meaning sea giant. For some people, including some who have lived and breathed the sea, it is far from a legend. Morgawr is a real living sea monster.

In July 1976, the head and neck of Morgawr startled the crew of a small inshore fishing boat off the Lizard, the southernmost point on the British mainland, and its brief appearances close to beaches and cliffs near by have been reported by many locals and holidaymakers.

The fishmermen had just taken on ice at Newlyn and were heading into the English Channel for a day's wreck-fishing about 25 miles south of the Lizard. Just after midday, George Vinnecombe looked out of the starboard window and called to his fellow fisherman John Cock, who was below in the fo'c'sle. He had seen an object on the water, about half a mile away. It looked to him like an upturned boat. Cock rushed up on deck and Vinnecombe steered the boat towards the object, expecting to find somebody in trouble. As they manoeuvred closer, Cock climbed on to the foredeck and he could see that it was not an upturned boat. They looked at each other and wondered what it was they had discovered.

They put the engine out of gear, switched off, and slowly drifted up towards the object. George Vinnecombe, who had been fishing these waters for over forty years and was familiar with giant baleen whales, pilot whales and all manner of marine animals, had no idea what it was. From about 9m (30ft) away, they could see a back with leathery, black, shiny skin devoid of scales, about 5.5 – 6.1m (18 – 20ft) long, and with three humps. It was a beautiful day and the sea was calm and, below the surface of the water, they could see a bulky body which they estimated to be several tonnes in weight.

The two fishermen were mesmerized and were just wondering what was going to happen next when, about one metre (3ft) in front of the body, a head and neck rose out about one metre (3ft) out of the water. The head, according to Vinnecombe, looked like that of a seal and had big eyes, while Cock thought it was about the size of that of a sheep. Then, without any commotion, the head, neck and body simply submerged back into the water. The fishing boat motored around for a while, searching for the creature with the echo-sounder fish-finder, but they saw it no more. Back at their home port, they were reluctant to talk about their brief encounter. Indeed, when they did speak of the incident to their fellow fishermen, their legs were pulled mercilessly, although one fisherman from Porthleven did reveal that he had seen a similar strange creature.

In the same year, the editor of *Cornish Life,* Dave Clarke, had an equally intriguing encounter with Morgawr a little further along the coast, not far from the port of Falmouth. There had been several sightings of the monster by holidaymakers and Clarke decided to put together an article for the magazine. He discovered from old newspapers that there had been many reports of a Cornish sea monster dating back over one hundred years. He interviewed several fishermen who had had similar experiences to George

Vinnecombe and John Cock, and then went to meet a local character, Tony 'Doc' Shiels, a self-acclaimed 'wizard', psychic, magician, monster-watcher and contributor to the *Fortean Times*, a periodical which specializes in reporting unusual phenomena.

During the interview, Shiels said that, with the help of his telepathic powers, he could summon Morgawr to the surface. Clarke was intrigued and very sceptical about the whole thing, but suggested that the sight of a wizard invoking the monster might make a good photograph to go with the article. They decided that Parson's Beach at the mouth of the Helford River was a likely spot, for there had been a sighting near by only a few months earlier.

On the morning of 17 November, they parked the car at Mawnan and, accompanied by Clarke's dog Sam, scrambled down the cliff-face between the trees and set up on the shingle beach. Shiels made several incantations and waved his arms about while Clarke took his photographs. Throughout the ceremony nothing appeared off the beach. Just as they were preparing to leave, however, Shiels noticed a small head sticking out of the water, about 90m (100yds) away. At first, Clarke thought it was a seal, as they are quite common in that area but, as the object came nearer and turned side-on, he could see it had an arched, slim, muscular neck that was quite uncharacteristic of a seal. It swam closer to the beach, to within 18m (60ft) of the two observers and swirled around in the water, causing considerable disturbance with what Clarke considered to be a very large body. Although it was difficult to give the creature an accurate size, he thought it was at least 15 – 18m (50 – 60ft) in length. The colour of the head, neck and the small section of body visible above the surface was dark grey-green.

He changed the standard lens on his camera for a telephoto lens and, looking through, he could see that the animal had a blunt nose and a rounded head with two 'buds' on the top. Both men took photographs feverishly, Clarke with his 35mm Pentax and Tony 'Doc' Shiels with a large-format Rolleiflex.

Dave Clarke was understandably suspicious about the whole event. Only the previous week he had seen Shiels saw a woman in half and bend spoons. Could he then have concocted the apparition as some kind of illusionary trick? But Tony 'Doc' Shiels seemed as shocked and surprised by the arrival of the creature as Clarke had been, although later, back at the pub and after a few pints, he did claim to have caused it to appear.

Meanwhile, Clarke's dog wanted its master to throw stones in the water and ran towards the water's edge, barking loudly. The creature then raised its head, opened its mouth, and sank slowly under the water. The entire episode lasted no more than half a minute. They stayed and watched for another hour but Morgawr did not reappear that day.

And the photographs? In the confusion, the wind-on mechanism of Clarke's camera had jammed and the film was only moving across at a half-a-frame at a time so the resulting pictures were overlapping. They had a little more luck with 'Doc' Shiels's 2¼ square camera. Despite the wide angle of the lens, they were able to blow up the colour pictures to show a rather indistinct image of the creature's head and neck.

Even more interesting photographs turned up mysteriously at the offices of the local newspaper, the *Falmouth Packet,* when an anonymous photographer who simply signed her name Mary F (Falmouth) submitted two photographs of Morgawr taken at Trefusis Point, not far from Falmouth Docks in Carrick Roads at the mouth of the River Fal. The newspaper published them on 5 March 1976, putting them on the front page under the heading 'Sir: kindly find enclosed one sea serpent', with a faked picture for comparison created by the newspaper's photographic reproduction staff, and with Mary F's accompanying letter. In the letter, dated 29 February 1976, the anonymous photographer wrote:

> The enclosed photos were taken by me about three weeks ago from Trefusis. They show one of the large sea creatures mentioned in your paper recently. I'm glad to know that other people have seen the giant brownish sea serpent.
>
> The pictures are not very clear because of the sun shining right into the camera and the haze on the water. Also I took them very quickly indeed. The animal was only up for a few seconds. I would say it was about fifteen to eighteen feet long. I mean the part showing above the water. It looked like an elephant waving its trunk, but the trunk was a long neck with a small head on the end, like a snake's head. It had humps on the back which moved in a funny way.
>
> The colour was black or very dark brown and the skin seemed to be like a sea-lion's.
>
> My brother developed the film. I didn't want to take it to the chemist. Perhaps you can make them clearer. As a matter of fact the animal frightened me. I would not like to see it any closer. I do not like the way it moved when swimming.
>
> You can put these pictures in the paper if you like. I don't want payment, and I don't want any name in the paper about this. I just think you should tell people about this animal. What is it?
>
> Yours sincerely
> Mary F
> (Falmouth)

The editor of the *Packet* and Peter Costello, author of *In Search of Lake Monsters,* tried to find Mary F, but only received a further letter which told how the negatives of the photographs had been sold to an 'American gentleman'.

The photographs and letter, however, prompted more correspondence. Miss M Jenkins recalled a friend telling her that she had seen a large snake-like animal swimming in the sea near Mylor Bridge several years before. And Miss Amelia Johnson, from West London, had been visiting her sister in Falmouth. She wrote:

> One afternoon we decided to take a walk. We parked the car by Mawnan church and, after having a brief look around the church, we started to walk in the direction of Rosemullion.
>
> My sister being quite fascinated by the old trees growing down the bank at the back of the church, I proceeded to walk on by myself.
>
> Looking out to sea, I saw a strange form suddenly emerge from the water in Falmouth Bay. It was just like the sort of description one hears of the Loch Ness Monster, a sort of prehistoric dinosaur thing, with a neck the length of a lamp-post.
>
> I raced back to my sister and told her what I had seen. She told me she thought I must be 'batty', and when we went back to see if we could see it again, it was no longer there.
>
> . . . The only point on which I disagree with Mary F is in the colouring. I should say it was more of a dark grey; but then, like her, I didn't stay around too long.
>
> In your introduction to the letter and photographs, you pointed out the possibility of faking such photographs. Since that is, of course, true, I thought under the circumstances I would take the opportunity of verifying Mary F's statement.

The following month, under banner headlines: 'Yna omma ow-tos an Morgawr' (Here comes the Sea-giant) the newspaper carried excellent photographs of a 'monster' in the Penryn River which seemed to follow very closely behind a local trawler. The red and green neck, which developed 'an ominous crick', and the three humps were spotted by the crew of a tanker moored at Coast Lines Wharf at about 0930hrs but were ignored by Falmouth Coastguards – after all, the date was 1 April.

And one curious thing about the Mary F photographs is that they resemble quite closely a small reproduction of a photograph, captioned 'The Great Worm of Lough Leane, Co. Kerry, Ireland', that appeared on page 29 of a publication called *Nnidnid: Surreality,* Number One, August 1986. The picture was taken by Tony 'Doc' Shiels.

The recent spate of Morgawr reports seemed to have started in September 1975, when two people from Falmouth, Mrs Scott and Mrs Riley, were at Pendennis Point. They watched a creature, which fits the description of the Parson's Beach monster, dive down and return to the surface grasping a writhing conger eel. It was, they said, a humpbacked creature with stumpy horns and bristles down its back.

In January 1976, Duncan Viner, a Truro dental technician, saw what he thought at first was a whale in the sea off Rosemullion Head. After a short while he saw a long neck appear at one end of the 9.1 – 12.2m (30 – 40ft) long dark shape. Gerald Bennet of Seworgan saw a similar creature in the Helford River.

In May, two bankers from London, Tony Rogers and John Chambers, were interviewed by the *Packet* about a creature that appeared before them while fishing. They were on the rocks near the infamous Parson's Beach, but were blissfully unaware of the monster stories. They were in for a surprise.

> Suddenly, something rose out of the water about 100 or 200 yards away. It was greeny-grey in colour and appeared to have humps. Another smaller one also appeared. They were visible for about ten seconds and looked straight at us.

Tony 'Doc' Shiels, saw the monster first on Sunday 4 July 1976, not long after the fishing boat incident. He and his family were on Grebe Beach, enjoying the unusually hot weather, when he thought he saw something in the water. Each time he brought the binoculars to bear, however, it could not be seen. Eventually everybody started to look and the children were the first to spot the unidentified creature. Shiels's wife wrote to the *Packet* about the experience. She told, for instance, of how she could only see the object if she looked out of the corner of her eye.

'For several seconds,' she wrote, 'I saw a large, dark, long-necked, hump-backed beast moving slowly through the water, then sinking beneath the surface.'

The Shiels family thought, at first, they had been hallucinating. It was, after all, the year of the great drought, with a long hot summer of extraordinarily dry weather. And, maybe others were hallucinating too, for there were numerous sightings reported throughout the summer and into the autumn.

Brother and sister holidaymakers from Gloucestershire, Allan and Sally White, were also on Grebe Beach on Sunday morning, 12 September, when they were startled by a 4.6 – 6.1m (15 – 20ft) long brown object that slithered off the beach and into the sea.

The Parson's Beach episode followed a couple of months later, and that was the last that was heard of Morgawr until four years had passed. On 20 February 1980, Geoff Watson, a sociology undergraduate from Thames Polytechnic, spotted an object 274m (300yds) from the shore near Helford Passage. Watson, a member of the Loch Ness Monster Association of Explorers, was patrolling the cliffs and beaches in the Mawnan and Trefusis areas in an attempt to get pictures of Morgawr. He managed to take a few shots of the dark object with a telephoto lens attached to his camera. He told the local *Western Morning News* that he had seen 'a series of protrusions on the surface giving the impression of being black'.

> The first protrusion rose about nine inches above the surface. Farther back there was what might have been a hump and then another hump at about equal distance.
>
> Of course, it might have been a log, but it certainly moved closer and closer to the shore and then disappeared.

Watson's sighting lasted no more than a minute, and when he later returned to London and processed the film, he found that the object was far too indistinct to be submitted as evidence.

Monster sightings around the Falmouth area have tailed off as of late, but there is a history of sea monster reports to consider. Morgawr is not, it seems, a new phenomenon. In 1876, the focus of attention was in Gerrans Bay, to the east of Falmouth. The report was in the *West Briton*.

> Portscatho. The sea serpent was caught alive in Gerrans Bay. Two of our fishermen were afloat overhauling their crab pots about 400 – 500 yards from the shore, when they discovered the serpent coiled about their floating cork. Upon their near approach it lifted its head and showed signs of defiance, upon which they struck it forcibly with an oar, which so far disabled it as to allow them to proceed with their work, after which they observed the serpent floating about near their boat. They pursued it, bringing it ashore yet alive for exhibition, soon after which it was killed on the rocks and most inconsiderately cast again into the sea.

How anybody can continue with their work after an encounter with a strange sea creature beats me. And the report does not say what happened to the creature after it had been 'cast again into the sea'. It was to appear, however – or at least a surviving relative – about fifty years later. Another report in the *West Briton* told the story of Mr Reese and Mr Gilbert, two fishermen who were trawling 5km (3mi) south of Falmouth and caught themselves a sea monster. It was 6m (20ft) long, had a 2.4m (8ft) tail, a beaked head, scaly legs, and its back was covered in matted brown hair. What happened to the body is not clear.

Other carcasses have appeared on Prah Sands in Mount's Bay in 1933 and on Durgan Beach at the mouth of the Helford river in January 1976. None of the animals was identified.

Further east along the coast, at the shark-fishing port of Looe, Harold Wilkins and a companion were at the waterside in 1949 when something very strange occurred. They spotted:

> Two remarkable Saurians, 19 – 20 feet long, with bottle-green heads, one behind the other, their middle parts under the water of the tidal creek of East Looe, Cornwall, apparently chasing a shoal of fish up the creek. What was amazing were their dorsal parts: rigid, serrated and like the old Chinese pictures of dragons. Gulls swooped down towards the one in the rear. These monsters – and two of us saw them – resembled the plesiosaurus of Mesozoic times.

This was quite an extraordinary sighting, for the creatures would have had to pass straight through the middle of a busy fishing port.

Leaving Morgawr and the other Cornish sea monsters, we next move across the Channel to the Channel Islands, for at Herm a sea monster was once spotted out of the water.

The area is known for the enormous tidal range and the speed with which the tide comes in and goes out. It is said that the sea at nearby St Malo on the French coast moves in and out faster than the speed of galloping horses. In these conditions a sea animal could easily find itself high and dry or at least trapped in a pool on the beach or among the rocks.

The year was 1923, and according to a letter sent by Mrs Hilda Bromley from Kensington, London, to Tim Dinsdale and published in his book *Loch Ness Monster,* a party of fourteen people, including guests of Sir Percival and Lady Perry, went to the beach at low tide to search for lobsters. The group arrived at a large pool and noticed marks on the seaweed as if a large creature had dragged across its

Archives: The University of Glasgow

Large sunfish are not commonly encountered at sea. The one *above* was exhibited in the Scottish Oceanographical Laboratory and was caught in the South Atlantic, to the northeast of the Falkland Islands. The bizarre shape has always intrigued those who see it (*below*).

Trustees of the National Museums of Scotland

Bobby Tulloch

A row of otters swimming in a line off the Shetland coast could easily be mistaken for a writhing, serpentine sea monster.

great bulk. They were all intrigued and started to follow the trail.

They came to a second, even larger, pool into which the trail disappeared. They stood at the edge and watched.

> Then slowly, away in the middle of the pool, a large head appeared and a huge neck – but we did not see the body; there it stayed with its great black eyes gazing at us without fear – then slowly it sank back into the water. It was evident that it had never seen a human before. We joined hands and all stepped into the pool, to see if we could disturb the creature, but it was too large and too deep for us to make any real impression.

Not far away, across the Golfe de Saint-Malo at Etables on the Brittany coast, a different sort of monster appeared before the Waymark family, of Tunbridge Wells. In a letter to Maurice Brown and Martin Chisholm, who had produced a BBC discussion, 'The Great Sea Serpent' broadcast in 1961, Mrs Waymark recalled the sighting in 1939:

> A huge creature like a serpent was swimming rapidly along on the surface of the water, weaving in and out and leaving quite a wake after it. We estimated it would be at least 20 feet long.

Back on the English side of the Channel, near Salcombe in South Devon, there was a similar strange serpent seen thirty-one years later, but with a new twist. Some thought it to be a conger eel, others suggested a basking shark, but the strange thing about this monster is that it made a noise.

The monster first made its presence known in August 1970 when skin-divers were about 12 – 24m (40 – 80ft) down in the sea off Lannacombe Bay. They heard a barking noise. At least four respected divers are known to have reported the strange and eerie sounds. The noise was so persistent throughout the next couple of months that it became known as the 'Lannacombe Bark'. Nothing, though, was seen below the surface.

At about the same time, the Salcombe Shark Angling Society reported that they were hooking something in mid-Channel, playing it for a while, and then losing contact. When the lines had been reeled in the anglers found that their large and formidable hooks had been bitten in two. They thought that even the strongest known shark to be found in British waters was incapable of doing such a thing.

In November, members of the Torbay branch of the British Sub-Aqua Club revealed that there was, indeed, a monster living at the bottom of the sea off Lannacombe, but that it was an easily recognized one – a giant conger eel. Estimates put its length at well over 2.4m (8ft), and the weight in excess of 54kg (120lb). This is not unusual, for skin-divers have reported seeing large female congers up to 2.7m (9ft) long. There was even one story in September 1987 of a 3.7m (12ft) long conger, nicknamed 'Elvis', in Haslar Lake at Gosport, Hampshire, which attacked, so it is claimed, a teenager on the foot. Afterwards, when the local authority drained the 13 acres of lake, all that could be found was a 0.6m (2ft) long common eel which was immediately dubbed 'Tiny Tim'.

Conger eels tend to stay near rocks or wrecks on the sea bottom and are rarely seen swimming at the surface. Nevertheless, during June of the following year reports began to accumulate that the head of a 'giant and fearsome' animal had been seen at the surface of the sea between Start Point and Lannacombe Bay. Locals and visitors, who had seen the creature, felt that the skin-divers' conger and the monster were two different creatures. The divers were also at a loss to explain away the noise.

It was not, however, the first appearance of a sea monster in this area. In 1906, and a few kilometres along the coast from Lannacombe, at Berry Head near Brixham, a buff-coloured 'baby' monster wriggled past a small sailing boat. It was estimated to be about 1.8m (6ft) in length, 13cm (5in) wide, and quite flat 'like a broadsword blade', according to yachtsman A J Butler. 'The edges were all serrated and gave the impression of being set along with tiny fins.' Was this, perhaps, a giant eel larva preparing itself for greater things? Or maybe it was a small oarfish, a none the less strange but well-documented fish that could easily be mistaken for a sea serpent? The description tallies with a fish commonly found on fishmongers' slabs in Sicily and known as the 'spadola', meaning cutlass. It is the scabbard fish which is fished from deep waters in the Straits of Messina in the Mediterranean and off Madeira in the Atlantic Ocean. Whatever it was, a larger version appeared six years later in the same area off Prawle Point.

On 5 July 1912, the captain and first mate of a German ship, the *Kaiserin Augusta Victoria,* witnessed a 6.1m (20ft) long, 46cm (18in) thick, eel-like creature thrashing about at the surface. It seemed to be fighting another sea animal. Its colour, which thcy could scc clearly, was blue-grey on the back and white below – a typical cryptic coloration pattern of a marine animal, whether it be eel, snake or marine mammal. (Seen from below, the white belly matches the light from the sky; seen from above, the grey-coloured back blends with the blue-grey darkness of the sea.)

In the same year a 'long sinuous body' was seen off a beach

further east along the English Channel at St Margaret's Bay in Kent. It was said to move rapidly and with undulations. And another 18.3m (60ft) long series of rapidly moving, pointed blobs were seen to submerge rapidly off the Norfolk coast at Kessingland Grange.

In 1925, a bulky, long-necked type of monster appeared in the Channel. An engineer on a small coaster, Mr H. Hodgson, was another eyewitness who was prompted to write to Maurice Brown after his radio programme. His letter was short and to the point.

> I listened with a certain amount of interest in your programme 'The Great Sea Serpent'. I have seen a plesiosaurus.

Further inquiries revealed that he, and many other of the crew, had seen a very large animal.

> It was of great bulk, as big as a 500-ton ship and with a very long neck and a small head . . . It had a terrific turn of speed.'

Mr Hodgson felt very privileged to have seen the creature.

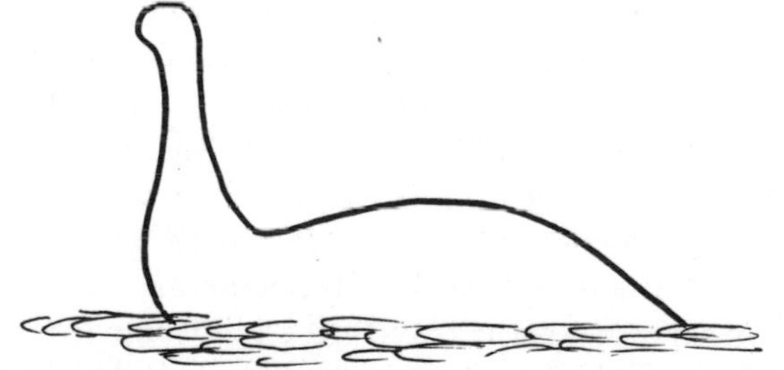

An English Channel sea serpent. *From a drawing by Mr H Hodgson.*

At the southern end of the North Sea, there is the mouth of the River Thames. It has always been a busy place with its docks and sea and river traffic. Nevertheless, sea monsters have not been discouraged from making an appearance. There was even a sea serpent in the Thames estuary itself, despite the river being an open sewer at the time. It was seen in an area appropriately known as the Black Deep, which had been closed to shipping during the First World War. The Royal Navy ship HMS *Kellett* went there in 1923 to survey the area and at about nine o'clock one bright August morning the Captain and navigator twice spotted a two-metres long serpentine neck sticking out of the water. It remained in view for no more than five seconds at a time, but, nonetheless, four years later the Captain was able to furnish Rupert Gould, author of *The Case for the Sea-Serpent*, with a rough sketch.

The Black Deep sea serpent of 1923. *From a drawing by Captain Haselfoot.*

Many years later, in 1950, Londoner John Handley was swimming in the sea near the Lido Baths at Margate, when he and a woman swimmer were more than a little surprised to see a 0.6m (2ft) long head, with a horse's ears, rise out of the water.

A little further to the north, off the Norfolk and Suffolk coasts, horse's-head monsters dominate reports during the 1930s. In 1930, a steward on a coaster saw a 1.2 – 1.8m (4 – 6ft) long head and neck 'like that of a camel'. The observer saw no ears. In 1931, another creature was seen near Thorpeness, and in 1933 *The Times* reported the fast-swimming humps and a head that swung from side to side of a creature watched by Mrs Sybil Armstrong and two of her staff.

In 1936, a former Norwich Lord Mayor, former Member of Parliament, and a sitting Member of Parliament were standing one evening on Eccles Beach when a 12.2m (40ft) sea serpent put in an appearance. It made worm-like movements, so the assembled dignitaries observed, and it travelled rapidly on the sea's surface, parallel to the shore. In 1938, two lesser mortals but undoubtedly characters who were better versed in the ways of sea and therefore less impressed by strange events – a couple of Southwold fishermen – were terrified by a 18.3m (60ft) long grey, humped creature. It had a big head and long neck that towered over them. The animal, probably more frightened of the fishermen than they of him, made off. It swam away at a speed of about 30 knots, like 'a torpedo'.

The Southwold sea serpent of 1938. *From a drawing by Ernest Watson and William Herrington.*

The stretch of Lincolnshire coast between Skegness and Mablethorpe has been a favourite place for monster-spotting. In the late 1930s, for example, Mr R W Midgley was going for an early morning stroll along the sea-wall at Trusthorpe (adjacent to Mablethorpe). Suddenly he noticed in the water what he described some time later in a letter to the *Skegness Standard* as 'four or five half-links of a partly submerged, huge snake-like body'.

About thirty years later, at Chapel St Leonards, the *Standard* reported another monster story. Sheffield steel-worker, George Ashton, was walking on the beach with his wife May, in October 1966, when he spotted something peculiar about 90m (100yds) out in the sea.

> It had a head like a serpent and six or seven pointed humps trailing behind. At first I thought it was a log, but it was travelling at about 8 mph and going parallel with the shore. We watched it for some time coming from the direction of Chapel Point until it disappeared out of sight towards Ingoldmells.
>
> I just didn't believe in these things and tried to convince myself that it was a flight of birds just above the water. But it was leaving a wake in the water . . . There was no noise. It just skimmed through the water.
>
> I have always been the first to laugh at the Loch Ness Monster . . . but not any more.
>
> I will swear on oath about what I saw.

Not far from the Ashtons was another middle-aged couple who were watching the event. George Ashton appealed to them, through a newspaper article, to come forward and verify his story. They never made contact, but a week later a Skegness man, John Hayes, wrote to the *Skegness Standard.* He had been cycling along in front of the Derbyshire Miners' Centre at Winthorpe at the beginning of the summer season. It was a clear, moonlit night. Suddenly, he was startled by a loud crack that came from the direction of the sea. He stopped and looked, and there in the water he could see a huge, dark shape moving at about 32kph (20mph) some 457m (500yds) from the shore.

The Chapel St Leonards stories also prompted a letter from a Spilsby resident, B M Baylis. He wrote:

> The description of a 'monster' seen in the sea off Chapel St Leonards is very reminiscent of one my friend and I saw off the Yorkshire coast some years ago.
>
> On a afternoon in early August, 1945, we were sitting on the

edge of the low mud cliffs at Hilston between Hornsea and Withernsea. There we saw a creature with a head and four or five rounded humps each of which was leaving a wake.

It was moving rapidly but quite silently along the shore northwards in face of a northerly wind.

Nobody at the time believed our report, but we are convinced we saw something.

Yorkshire, too, has had its own share of sea monsters. At Filey, half way between Scarborough and Flamborough Head, in 1934, fishermen had seen a strange beast about 4.8km (3mi) from the coast. Then, one very dark, moonless night, according to a report in the *Daily Telegraph,* Filey coastguard Wilkinson Herbert was walking along Filey Brigg, a mile-long, thin bill of rocks that protects Filey Beach from north-east winds, when he came upon something quite extraordinary.

Suddenly I heard a growling like a dozen dogs ahead; walking nearer I switched on my torch, and was confronted by a huge neck, six yards ahead of me, rearing up 8 feet high!

The head was a startling sight – huge tortoise eyes, like saucers, glaring at me, the creature's mouth was a foot wide and its neck would be a yard round.

The monster appeared as startled as I was. Shining my torch along the ground, I saw a body about 30 feet long. I thought 'this is no place for me' and from a distance I threw stones at the creature. It moved away growling fiercely, and I saw the huge black body had two humps on it and four short legs with huge flappers on them. I could not see any tail. It moved quickly, rolling from side to side, and went into the sea. From the cliff top I looked down and saw two eyes like torchlights shining out to sea 300 yards away. It was a most gruesome and thrilling experience. I have seen big animals abroad but nothing like this.

Easington, where North Sea Gas is today purified for the national grid, was the site of another monster sighting in the late 1930s. The experience only came to light after the 1961 BBC radio programme. Joan Borgeest was teased by her friends when she told of her experience and eventually thought it better not to mention it to anybody. The programme prompted her to write about her unusual and inexplicable sighting.

The event took place when the long, sandy beach was relatively empty. Mrs Borgeest was looking after some children and idly looking out to the North Sea.

> Suddenly I saw a huge creature rise; it was a green colour, with a flat head, protruding eyes, and a long flat mouth which opened and shut as it breathed; it was a great length and moved along with a humped glide.

Mrs Borgeest called to some people nearby but this seemed to scare the animal, which was only 90m (100yds) away, and it dived and did not reappear.

The Eassington sea serpent. *From a drawing by Mrs Borgheest.*

Across the border, the first ever report in a British newspaper of a sea serpent to be seen off the North Sea coast of Scotland, appeared in *The Scotsman* on 19 November 1873, and was included in an extract from a letter from Mr Joass, of Golspie, to the Reverend John Macrae, of Glenelg (another sea monster witness as you will read later). It was published in the December 1873 edition of *The Zoologist*.

> On Tuesday afternoon last, Lady Florence Leveson Gower and the Hon Mrs Coke, driving near the sea, about eight miles east from Dunrobin, saw what seemed to them a large and long marine animal. On Wednesday morning Dr Soutar, of Golspie, saw a large creature rushing about in the sea, about fifty yards from the shore: it frequently raised what seemed a neck, seven feet out of the water, and from the length of troubled water behind it appeared to be fifty or sixty feet long. He said to his family on meeting them at breakfast, 'If I believed in sea serpents, I should say I had seen one this morning.' I may mention that this gentleman is a most trustworthy observer and cautious man.
>
> On Thursday I saw what seemed some drift seaweed. When your report was published Dr Taylor, the author of *Thanatophidia of India,* was at the castle; I asked him what he thought of the matter, and he said he was quite prepared to believe in such a monster. Mr Vernon Harcourt told me that he was in a small yacht off Glenelg on the evening of the day mentioned in your report, and about six miles from the locality, and that he and his crew saw what seemed a great moving mass, which, but for some engagement or the lateness of the hour, they would have examined.

The sightings started in mid-September and the day after *The*

Scotsman article, on 20 November, *The Times* included another report from Mr Joass, this time stating that he was an eyewitness and drawing attention to appendages on the creature's head.

> The ears seemed to be diaphanous and nearly semicircular flaps or valves overarching the nostrils, which were in front. The cavity of the eye appeared to be considerably further back, and a peculiar glimmer in it, along with the sudden disappearance of the creature, presented, indeed, the only signs of its vitality, so far as I could see, while I watched it for half an hour apparently drifting with the rising tide, but always keeping the same distance off shore.

After Mr Joass's letter had reached the pages of *The Times,* considerable correspondence followed in which the 'monster' was attributed to a school of porpoises, a sea turtle, an oarfish, a log, a basking shark and a flock of birds.

The Dunrobin sea serpent of 1873. *From a drawing by the Revd James Joass.*

But, despite being well and truly rejected, the creature reared its head again, further south, on 18 November 1873 (significantly the day *before* the Dunrobin article was first published). A 'long and large black animal', which moved back and forth showing a head and undulations, was watched by a crowd of over one hundred people at Belhaven Bay, near Dunbar at the mouth of the Firth of Forth. The animal was about 400m (¼ mi) offshore. This was part of the report that appeared in *The Scotsman* two days later:

> Sometimes it appeared to stretch itself out to its full length, at which times both its head and tail were seen above the water . . . Most frequently, however, it was the undulations or apparent coils of the body that were observed, two or three of them being occasionally visible at the same time . . . it seemed to be upward to a hundred feet in length, with an apparent breadth of from two to three feet.

The crowd was able to witness the performance of this incredible creature for over a quarter of an hour, in which time it surfaced and dived every couple of minutes.

The Firth of Forth was again in the news many years later, in July 1939, when a small girl saw an animal, not more than 366m (400 yards) from the shore, at Dunbar. She told her father, and when they both returned to the beach they were able to watch it for over an hour and a half. On the same day, some fishermen at West Wemyss, on the northern shore of the Firth of Forth, saw a large brown creature with a horse's head and large prominent eyes.

It was also fishermen who, in October 1892, witnessed the 12.2m (40ft) long snake-like animal near Broughty Ferry at the mouth of the Tay estuary, a little to the north of the Firth of Forth. Skipper Thomas Gall and his crew aboard the *Catherine* first saw what looked like a dark-blue-coloured snake's tail. As it disappeared, a large blackish-brown head popped up just a few metres from the boat. A couple of years later, father and son from the Birnie family saw a long snake-like thing on the beach at New Aberdour.

In 1903, sea serpent hysteria seemed to hit the east coast of Scotland once again when the fishing boat *Rosa* encountered a 1.8m (6ft) head and neck 16km (10mi) south-east of Montrose, while near Fraserburgh the trawler *Glengrant* was reputed to have been 'attacked' by a 60m (200ft) long sea monster with a head much like that of a sea-horse, large glistening green eyes, an enormous mouth filled with evil-looking teeth, and a long fin down the back. The eyewitness account reads like a horror story:

> When 20ft off the vessel it reared to a great height above, and with a loud hissing noise plunged down again . . . The vessel took a great dip by the bow and shipped a huge sea, washing the deck clear and flooding the engine-room, cabin and fo'c'sle . . . To the horror of the crew it was seen turning and coming in again at a furious pace.

One crewman fetched a rifle and shot at the monster which dived immediately. Its body, according to the eyewitnesses, was long and sinuous and wriggled like a snake. The national Press had a field day with outrageous tales of unlikely monsters.

Across the often treacherous waters of the Pentland Firth at the northernmost tip of Scotland are the Orkney Islands. There are about seventy islands and islets of which about eighteen are inhabited. They have not only played host to the carcass of the Stronsa beast (see page 180) but also to living monsters.

In 1910, for example, three wildfowlers, W.J. Hutchinson of Kirkwall, Orkney, his father and his cousin, were heading in their sail boat for the Skerries of Work to shoot duck and plover. They were surprised to see a school of whales leaping clear of the water and moving out of the area at high speed. The occupants of the boat were concerned that the whales would turn and crash into their fragile vessel. In a letter to Tim Dinsdale, which is published in his book *Loch Ness Monster*, Hutchinson recalled what happened next.

> My father was steering, and after the whales had disappeared, looked ahead on the course for the Skerries – and then I heard him say, 'My God boys, what's that?' pointing ahead. I looked up and saw a creature standing straight up out of the sea – with a snake-like neck and head, like a horse or camel!

The boat was turned so as to be beam-on to the strange creature and Hutchinson went for his gun. His father was concerned that they might injure the animal and it would sink the boat. The animal, though, eventually sank slowly into the water until, without any bubbles or commotion, it disappeared.

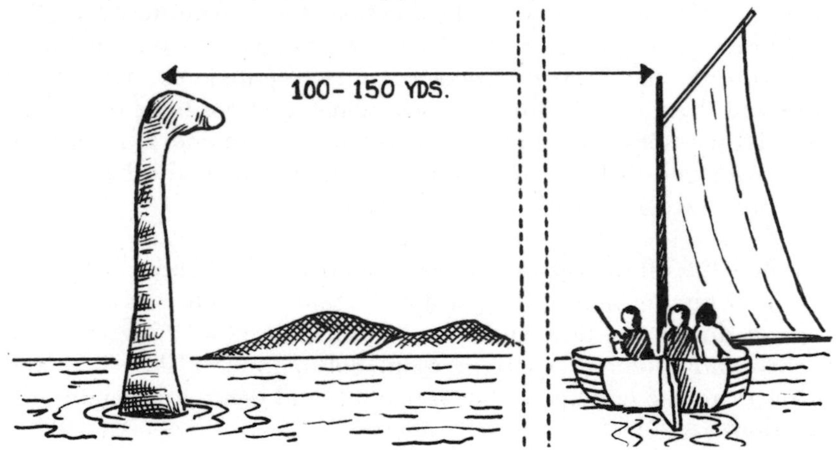

The Meil Bay sea serpent of 1910. *From a drawing by W J Hutchinson.*

Nine years later, lawyer J MacKintosh Bell, writer to the Signet, of Roundstonefoot, Moffat, wrote to Rupert Gould telling of an encounter with a strange sea creature in the Pentland Firth between the Scottish mainland and the Orkneys. MacKintosh Bell was on holiday on Hoy and helping the crew of a local fishing boat. Early one morning in August they were picking up lobster pots between Brims Ness on the mainland and Tor Ness at the southern end of the island and, as the crew were chatting, they mentioned a curious animal that they had seen several times in that area. Right on cue the creature appeared. They saw a neck the thickness of an elephant's leg on top of which was a small dog-like head. It had small black eyes and whiskers. Bell tried to take a picture of it but, as is common in these emotionally charged moments, his camera's shutter didn't work. Then he suggested taking a pot-shot with a rifle, but the skipper was afraid that if the animal was wounded, it might attack them. Finally, having failed to snap or shoot it, he resorted to sketching it – the head and neck, and the overall shape of the body as it swam, about 3.1m (10ft) down, right under the boat. For some time afterwards, the creature surfaced near to the boats of Bell's fishing friends and many other lobster and cod fishermen, and it became a familiar topic of local gossip.

The Hoy sea serpent of 1919. *From drawings by J Mackintosh Bell.*

In Scapa Flow, to the east of Hoy, commercial diver Mr Faithful had an upsetting experience. He was in one of the salvage parties raising German ships from the floor of Scapa Flow. On one occassion, according to a letter sent to Tim Dinsdale, the salvage party dropped some depth charges to loosen part of a sunken warship and Mr Faithful followed down afterwards to assess the state of the wreckage. As he reached the bottom, he landed on something soft. It was the body of a large animal. It had a horse's head and was busily eating the fishes that had been killed or stunned by the explosion. A startled Mr Faithful gave the emergency three pulls on his safety line and was whisked back to the surface where the rest of the crew laughed at his story. The accuracy of the story is suspect in that it was not first hand. The letter was from W.J. Hutchinson - yes, the same W.J. Hutchinson who saw the Meil Bay Monster. He had heard the story from his sister, who, in turn, had heard it from a friend. Nevertheless, it's a good tale worthy of yet another telling.

Back on the Scottish mainland, the west coast village of Scourie became the focus of sea monster attention in 1851. Sir William Flower, the Director of the British Museum (Natural History) had been quoted in the national press as not rejecting the notion of sea serpents living in the world's oceans. He had, though, asked for some tangible evidence. 'Bring me a scale,' he had said, and Dr W M Russell did, or at least *tried* to do just that. The Scottish doctor wrote to *The Times:*

> I happened to be on a visit to the house of Mrs M'Iver . . . and I not only heard from her a detailed account of the recent appearance of a great sea serpent in the little bay of Greiss, opposite her residence, but I actually became the possessor of a number of scales about the size and shape of a scallop shell, which were found on the reef of rock in the bays whereon the monster had comforted itself by scratching its head. Mrs M'Iver had not merely heard of the strange creature: she had seen it taking a leisurely swim along the beach, to the great alarm of the fish, shoals of which leaped out of the water in front of it.

The monster came to lose its scales on the rocks after some local fishermen had done what humankind always seems to do in response to an unusual creature – they shot at it and wounded it. The animal rested for a while on nearby rocks until chased away by the approaching boats. One of the fishermen gave Mrs M'Iver some of the scales and she, in turn, passed them to Dr Russell. He, however, managed to lose them!

Turning to the islands off the west coast of Scotland, the Outer Hebrides have had their fair share of monsters. On 31 May 1882, the German ship *Katie,* returning from New York, was steaming past the northern tip of Lewis, about 12.8km (8mi) offshore. The captain and crew saw what they took to be wreckage in the water off the starboard bow. It looked like the upturned hull of a boat, with as much lying below the water as was visible above. The ship changed course to avoid the object. The captain noted that there were a series of bumps in the water that looked like rocks.

> When we changed our course obliquely from the object, which lay quite still all thc timc, to our astonishment there rose, about eight feet from the visible end, a fin about ten feet in height, which moved a few times, while the body gradually sank below the surface. In consequence of this the most elevated end rose, and could distinctly be made out as the tail of a fish of immense dimensions.
>
> The length of the visible part of this animal which had not the

> least resemblance to the back of a whale, measured, according to our estimation, about 150 feet; the bumps, which were from three to four feet in height, and about six to seven feet distant from each other, were smaller on the tail end than the head end, which withdrew from our observation.

The captain, Captain Weisz of the Stettin Lloyd shipping line, commissioned the American animal painter, Andrew Shultz, to make an engraving of the monster which, according to Weisz, had been seen by many Lewis fishermen.

And there were many more monsters to come. In January 1895, an 18.3m (60ft) long serpent with two 3.1m (10ft) coils showing above the water was seen by Angus Macdonald off the small island of Berneray at the southern end of Lewis. In February, a Free Kirk minister was at the other end of the island – the Butt of Lewis – when a giraffe-like neck with a ruffle 0.6m (2ft) behind the ears rose 4.6m (15ft) out of the water.

> It had two great staring eyes, like a bull's, fixed upon me, and I then saw three joints of its body 120 feet long fitting into each other like a lobster's tail.

A year later, Ivan MacGregor spotted a pair of them swimming in the sea between Lewis and Cape Wrath on the mainland.

The sea around the Inner Hebrides, particularly to the south and east of the island of Skye, is another favourite monster haunt. The first record was in June 1808 and it came from the Reverend Donald Maclean in a letter to Patrick Neill, the Secretary of the Wernarian Natural History Society of Edinburgh. Maclean was rowing off Coll, one of the small isles to the south of Rhum and east of Mull, when he saw what he thought was a small rock in a place where he had never seen a rock before. The rock, however, raised itself up and Maclean thought he saw an eye; all this at a distance of about 800m (½ mi).

> Alarmed at the unusual appearance and magnitude of the animal, I steered so as to be at no great distance from the shore. When nearly in a line betwixt it and the shore, the monster directed its head (which still continued above the water) towards us, plunged violently under water. Certain that he was in chase of us, we plied hard to get ashore. Just as we leaped out on a rock, taking station as high as we conveniently could, we saw it coming rapidly under water towards the stern of our boat. When within a few yards of the boat, finding the water shallow, it raised its monstrous head

above water, and by a winding course got, with apparent difficulty, clear of the creek where our boat lay, and where the monster seemed in danger of being embayed. It continued to move off, with its head above the water, and with the wind, for about half a mile, before we lost sight of it.

Its head was rather broad, of a form somewhat oval. Its neck somewhat smaller. Its shoulders, if I can so term them, considerably broader, and thence it tapered towards the tail, which last it kept pretty low in the water, so that a view of it could not be taken so distinctly as I wished. It had no fin that I could perceive, and seemed to me to move progressively by undulation up and down. Its length I believed to be from 70 to 80 feet.

The interesting thing to note about this sighting is that it is one of the first references to vertical undulations of the body, a method of progression which throws up some unusual anatomical questions and which we shall hear more about in later reports. Also, Maclean adds in his letter, he was not the only person to see the beast. There were encounters further north.

About the time I saw it, it was seen about the Isle of Canna. The crews of thirteen fishing boats, I am told, were so much terrified at its appearance, that they in a body fled from it to the nearest creek.

And there, for the next sixty-four years, the matter rested. But on 20 August 1872 that was all to change.

The Reverend John Macrae, Minister of Glenelg, his daughters Forbes and Katie, grandson Gilbert Bogle, the Reverend David Twopenny, Vicar of Stockbury in Kent, and an anonymous Highland lad were in a small cutter, called the *Leda,* bound for Loch Hourn. They had left Glenelg and were heading south-west along the Sound of Sleat. In the May 1873 edition of *The Zoologist* Macrae and Twopenny told their story.

It was calm and sunshiny, not a breath of air, and the sea perfectly smooth. As we were getting the cutter along with oars we perceived a dark mass about two hundred yards astern of us, to the north. While we were looking at it with our glasses (we had three on board) another similar black lump rose to the left of the first, leaving an interval between; then another and another followed, all in regular order. We did not doubt its being one living creature: it moved slowly across our wake, and disappeared. Presently the first mass, which was evidently the

> head, reappeared, and was followed by the rising of the other black lumps, as before. Sometimes three appeared, sometimes four, five, or six, and then sank again. When they rose, the head appeared first, if it had been down, and the lumps rose after it in regular order, beginning always with that next the head, and rising gently; but when they sank altogether rather abruptly, sometimes leaving the head visible. It gave the impression of a creature crooking up its back to sun itself. There was no appearance of undulation: when the lumps sank, other lumps did not rise in the intervals between them.

The two authors described seven humps, making eight with the head, and thought the head flatter than the other humps; a nose was visible above the water. They estimated the total length to be about 13.7m (45ft).

> Presently, as we were watching the creature, it began to approach us rapidly, causing great agitation in the sea. Nearly the whole of the body, if not all of it, had now disappeared, and the head advanced at a great rate in the midst of a shower of fine spray, which was evidently raised in some way by the quick movement of the animal – it did not appear how – and not by spouting.

Forbes Macrae was terrified and made a hasty retreat to the cabin and insisted, it seems, on being landed in the middle of the following night only to trek home alone some thirteen miles, over mountain paths by moonlight. The monster, in the meantime, had come to within 90m (100yds) of the boat and submerged. They could see the wake moving across the Sound and the head emerged from the water occasionally. Gilbert Bogle saw that the colour was dark slaty-brown and he described a small dorsal fin behind the head. Katie Macrae also saw a triangular fin she estimated to be the size of the boat's jib. She considered it was most likely a pectoral fin that could only be seen when the creature turned over.

The following day, the party, minus Miss Forbes, set out for the return home but were becalmed again on the north side of the mouth of Loch Hourn, an inlet on the eastern side of the Sound of Sleat. The monster was there waiting for them.

> As we were dragging slowly along in the afternoon the creature again appeared over towards the south side, at a greater distance than we saw it the first day. It now showed itself in three or four rather long lines and looked considerably longer than it did the day before: as nearly as we could compute, it looked at least sixty

> feet in length. Soon it began careering about, showing but a small part of itself, as on the day before, and appeared to be going up Loch Hourn.

Unbeknown to those sitting in the *Leda,* the monster had also been spotted by people from the island of Eigg 32.2km (20mi), south-west of the Loch. Later in the day, the boat had moved very little because of the lack of wind. They had nearly reached the island of Sandaig when the monster was seen again.

> It came rushing past us about a hundred and fifty yards to the south, on its return from Loch Hourn. It went with great rapidity, its black head only being visible through the clear sea, followed by a long trail of agitated water. As it shot along, the noise of its rush through the water could be distinctly heard on board. There were no organs of motion to be seen, nor was there any shower of spray as on the day before, but merely such a commotion in the sea as its quick passage might be expected to make. Its progress was equable and smooth, like that of a log towed rapidly.

It was also spotted by the skipper of another boat in the Loch entrance, one Finlay Macrae from Bundaloch, in the parish of Kintail, and by a lady at Duisdale on the south-east corner of Skye whose house overlooked the part of the sound which is opposite the entrance to Loch Hourn. She described seeing something that looked like eight seals in a row.

As the *Leda* slowly progressed northwards up the Sound of Sleat, the monster showed itself several times. And, at the point where the Sound of Sleat narrows, the ferrymen on either side of Kylerhea saw and heard it pass that same evening.

The following day Lord Macdonald and his guests aboard his steam yacht saw the monster at Loch Hourn, and the day after that boat-builder Alexander Macmillan from Dornie and his brother Farquhar were fishing in the entrance of Lochduich to the north-east of the narrows at Kylerhea, half-way between Druidag and Castledonan, when the head and four humps swam rapidly past. They heard the rush of water and saw the creature and its wake several times in the Loch. Each time they headed rapidly for the shore in case it should attack them. Macmillan estimated its length to be 18 – 24m (60 – 80ft).

David Twopenny added a postcript to the report, saying that he and Macrae had scarcely expected anybody to believe them. Nevertheless they felt that they should record the event for further discussion. Twopenny wrote:

> The animal will probably turn up on those coasts again, and it will be always in that 'dead season', so convenient to editors of newspapers, for it is never seen but in the still warm days of summer or early autumn. There is a considerable probability that it has visited the same coasts before. In the summer of 1871 some large creature was seen for some time rushing about in Lochduich, but it did not show itself sufficiently for anyone to ascertain what it was. Also some years back a well-known gentleman of the west coast, now living, was crossing the Sound of Mull, from Mull to the mainland, 'on a very calm afternoon, when,' as he writes, 'our attention was attracted to a monster which had come to the surface not more than fifty yards from our boat. It rose without causing the slightest noise, and floated for some time on the surface, but without exhibiting its head or tail, showing only the ridge of the back, which was not that of a whale, or any other sea animal that I had ever seen. The back appeared sharp and ridge-like, and in colour very dark, indeed black, or almost so. It rested quietly for a few minutes, and then dropped quietly down into the deep, without causing the slightest agitation. I should say that above forty feet of it, certainly not less, appeared on the surface.'

Twopenny went on to mention the fact that the people who live and sail these waters are familiar with whales, dolphins and seals. He also quoted Mr Maclean's encounter off the island of Coll in 1808 (see above) and then drew attention to the speed at which their own monster had been travelling. He thought that it would be the major stumbling-block in getting people to believe their story.

> On the 20th, while we were becalmed in the mouth of Loch Hourn, a steam launch slowly passed us, and, as we watched, we reckoned its rate at five or six miles an hour. When the animal rushed past us on the next day at about the same distance, and we were again becalmed nearly in the same place, we agreed that it went quite twice as fast as the steamer, and we thought that its rate could not be less than ten or twelve miles an hour.

Twopenny concluded:

> In the meantime, as the public will most probably be dubious about quickly giving credit to our account, the following explanations are open to them, all of which have been proposed to

> me, viz.: – porpoises, lumps of seaweed, empty herring-barrels, bladders, logs of wood, waves of the sea, and inflated pig-skins; but as all these theories present to our minds greater difficulties than the existence of the animal itself, we feel obliged to decline them.

The monster was not seen again until 1893 and reported in an article in *Strand Magazine* two years later. Kintail-born ear, nose and throat specialist Dr Farquhar Matheson, who practised in Soho Square, London, was accompanied by his wife whilst sailing in the Kyle of Lochalsh. This is a narrow channel separating the mainland from the island of Skye a little to the north of the places where the 1872 sightings took place. It was a fine, clear day. The sailing was good. Then, at between one and two in the afternoon, a long, thin, straight neck as tall as the boat's mast appeared about 183m (200yds) ahead. It was swimming rapidly towards the sailing boat.

The Lochalsh sea serpent of 1893. *From a drawing by Dr Farquhar Matheson.*

> Then it began to draw its neck down, and I saw clearly that it was a large sea monster – of the saurian type, I should think. It was brown in colour, shining, and with a sort of ruffle at the junction of the head and neck. I can think of nothing to which to compare it so well as the head and neck of a giraffe, only the neck was much longer, and the head was not set upon the neck like that of a giraffe; that is, it was not so much at right angles to it as a continuation of it in the same line. It moved its head from side to side, and I saw the reflection of the light from its wet skin . . . I saw no body – only a ripple of water where the line of the body should be. I should judge, however, that there must have been a

> large base of body to support the neck. It was not a sea serpent, but a much larger and more substantial beast – something of the nature of a gigantic lizard, I should think. An eel could not lift up its body like that, nor could a snake.

The creature appeared not to have scales but had smooth skin. It appeared three more times at intervals of two to three minutes, lifting its head vertically out of the water each time. The good doctor attempted to keep pace with the creature but, after about a mile, it disappeared into the distance.

In the early 1900s, a similar creature was seen by fishermen off Skye and Rhum. The stories were told by fisherman Sandy Campbell to basking shark fisherman and author Gavin Maxwell, and he included them in his book *Harpoon at a Venture.*

In the first encounter in Loch Scavaig at the southern end of Skye, Sandy Campbell, his uncle, and John Stewart were fishing for the herring that used to come into the loch in their thousands.

> It was dusk; the sky was still light, but the land was dark – a fine night with a light northerly breeze and a ripple on the water. Sandy and the two old men began to haul their net. He was only a young boy, and his arms tired easily; he rested for a moment, and as he did so he noticed an object rising out of the water about fifty yards to seaward of them. It was about a yard high when he first saw it, but, as he watched, it rose slowly from the surface to a height of twenty or more feet – a tapering column that moved to and fro in the air. Sandy called excitedly to the old men, but at first got only an angry retort to keep on hauling the net and not be wasting time. At last Stewart looked up in exasperation, and then sprang to his feet in bewildered astonishment, as he too saw what Sandy was looking at. While this 'tail' was still waving in the air they could see the water rippling against a dark mass below it which was just breaking the surface, and which they presumed to be the animal's body. The high column descended slowly into the sea as it had risen; and as the last of it submerged the boat began to rock on a commotion of water like the wake of a passing steamer.

The three occupants of the skiff (a small light boat used for inshore fishing) were understandably terrified. They dropped their nets and rowed furiously for the safety of the shore.

A few years later (Maxwell believes it could have been in 1917) Ronald Macdonald, his brother Harry, and their father were in a boat at the mouth of Loch Brittle when a similar kind of phenomenon presented itself to the three astonished fishermen. It was heading north at a speed of about five knots and about a mile away.

> It appeared as a high column, said to be a great deal higher than the object Sandy had seen in Loch Scavaig, and light flashed at the top of the column as though a small head were being turned from side to side. There was a considerable commotion in the water astern of it, but no other portion of the body was visible above the surface. It submerged slowly until nothing was left showing above the sea, and it seemed to descend vertically and without flexion.

The encounter, however, that gave rise to the most detailed account took place in September 1959 near the island of Soay, at the southern end of Skye. On the morning of the 13th, basking shark fisherman Tex Geddes (who once crewed with Gavin Maxwell) and engineer James Gavin, who was on holiday on the island, were out mackerel fishing. Visibility was good. They had seen a basking shark and a pod of killer whales in the distance, but their attention was drawn to a black shape about 3km (2mi) away across Soay Sound. It approached at a speed of about 3 – 4 knots and, as it came closer, they could hear it breathing. Geddes recalled the encounter in a letter to Dr Maurice Burton, the naturalist and writer who has written extensively about sea and lake monsters. The substance of the letter was published on 4 June 1960 in an article in the *Illustrated London News*.

> The head was definitely reptilian, about two feet six high with large protruding eyes. There were no visible nasal organs, but a large red gash of a mouth which seemed to cut the head in half and which appeared to have distinct lips. There was at least two feet of clear water behind the neck, less than a foot of which we could see, and the creature's back which rose sharply to its highest point some three to four feet out of the water and fell away gradually towards the after end. I would say we saw 8 to 10 feet of back on the waterline.

It came to within 18m (20yds) of the dinghy and was constantly turning its head from side to side. The two petrified occupants noticed that the head was blunt, there were no teeth, the body was scaly, and the mid-line of the back came to a knife-edged ridge which was deeply serrated. The creature seemed to breathe through its mouth, which opened and closed regularly. Geddes thought it likely to weigh above five tons. They watched, as it travelled in a SSW direction towards the island of Barra at the southern tip of the Outer Hebrides.

Dr Burton also had a letter from James Gavin:

The creature seen by Tex Geddes and James Gavin at Soay in 1959. Top: the sketches made by Geddes (left) and Gavin (right). Bottom: The reconstruction which appeared in *The Illustrated London News*.

Herewith, as briefly as possible, my description of the beast. At the waterline the body was 6 to 8 ft long. It was humped shaped, rising to a centrally placed apex about two feet high. The line of the back was formed by a series of triangular-shaped spines, the largest at the apex and reducing in size to the waterline. The spines appeared to be solid and immobile – they did not resemble fins. I only got a lateral view of the animal but my impression was that the cross-section of the body was roughly angular in shape. Apart from the forward glide I saw no movement.

The neck appeared to be cylindrical and, at a guess, about 8in. in diameter. It arose from the water about 12in. forward of the body. I could not see where they joined; about 15 to 18in. of neck was visible. The head was rather like that of a tortoise with a snake-like flattened cranium running forward to a rounded face. Relatively it was as big as the head of a donkey. I saw one

> laterally placed eye, large and round like that of a cow. When the mouth was opened I got the impression of large blubbery tendril-like growths hanging from the palate. Head and neck arose to a height of about two feet. At intervals the head and neck went forward and submerged. They would then re-emerge, the large gaping mouth would open (giving the impression of a large melon with a quarter removed) and there would be a series of very loud roaring whistling noises as it breathed. After about five minutes the beast submerged with a forward diving motion – I thought I saw something follow the body down. It later resurfaced about a quarter of a mile further out to sea and I then watched it until it disappeared in the distance. I have heard that the crews of two lobster boats, fishing north of Mallaig, have also seen this animal – much to their consternation.

Dr Burton went on to analyse the reports and the drawings. There was, he thought, an uncanny resemblance to a sea turtle, an animal that some biologists believe could account for many sightings of unusual and unidentified marine creatures in British waters.

A little to the south of this acclaimed sea monster area lies the sea loch – Loch Linnhe – that joins Loch Ness via a system of rivers and canals, site of the famous lake monster. Could the sea serpents seen off the Scottish isles be related to the legendary beastie or even be 'Nessie' on migration?

Eric Robbins, an ironmonger from Doncaster, wrote to the producers of the 1961 BBC Radio programme telling them how he saw a creature resembling a traditional Loch Ness monster on two occasions: once on 21 July 1954 in the seawater loch, Loch Linnhe, and again on 15 July 1960 in the freshwater loch, Loch Lochy (the expanse of water between Loch Linnhe and Loch Ness).

In Loch Linnhe, the island of Shuna was visited by a sea monster on the morning of 30 July 1887. Two naturalists, Professor Matthew Forster Heddle, and J A Harvey Brown, were cruising in the yacht *Shiantelle* when breakfast was rudely interrupted by the appearance of a monster just 457m (500yds) away from the boat. Two of the crew were also witness to the strange event. After the creature had disappeared each of the four men, without conferring, wrote down a description of what he had seen. Several days later they opened the sealed envelopes and compared notes. The descriptions tallied, but their interpretations did not. Professor Heddle wrote of a low flat head, much like a large 1.37m (4½ft) skate's, and a series of ten humps that increased in size towards the middle. The overall length was about 18 – 20m (60-65ft). Heddle felt that the entire body was rushing through the water.

> My impression was that, setting aside the quiescent low head, I did not see a solid substance at all – except when the tail hummocks momentarily appeared – and that what I did see was water being thrown over laterally by the undulous lashings of a long black fin of a dark colour, which gave opacity.

Brown agreed with most of the descriptions of hummocks and water movement, but was more inclined to focus on the 'lack of solid substance' mentioned by Heddle. Brown wrote:

> I came to the conclusion, and feel very certain still, that it was simply a tide-rip or tidal wave coming from the direction of Corrievreachan between Scarba and Jura running Easterly and then N. Easterly along smooth water where soundings showed the meeting of the shallow and the deep.

So much for the sea serpent. Brown and Heddle agreed to differ on the interpretations.

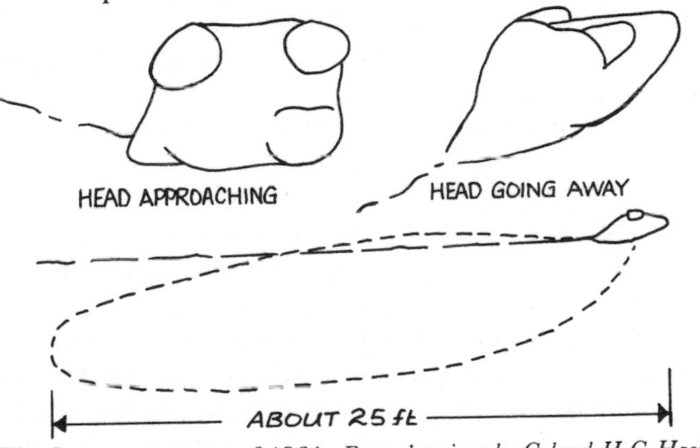

The Jura sea serpent of 1964. *From drawings by Colonel H G Hasler.*

Jura itself featured in a sea serpent record in early June 1964. Neil MacInnes, a stalker from Craighouse, was driving along near the shore and saw a creature that, in the distance, looked much like a box. When he drew closer he could see what appeared to be a cow's head. He took his telescope and could see that the head was, indeed, cow-like, but with white buds on either side of the head rather than horns. It moved from side to side. There were no visible eyes or ears, fins or humps. Although MacInnes could not see all of the body above the water, he estimated it to be about 7.6m (25ft) long, tapering at the tail, coloured grey, of smooth texture, and it moved faster than the slow tide flow, without wake or disturbance.

The island of Arran played host to a sea monster on 28 July 1931. It was evening when Glaswegian doctor John Paton and his 14-year-old daughter were cycling back to their holiday home. They stopped to investigate what they thought was an upturned boat on some rocks just off shore. Imagine their surprise when the bow of the boat turned around and looked at them. It had a head the shape of a parrot's, and its body was longer than a large elephant's. Had they seen an elephant seal or another seal-like creature? Dr Paton thought the creature did resemble an elephant seal but these animals only live in the Antarctic or along the Pacific coast of North America. The parrot-shaped head, though, suggests a sea turtle.

Three years later, Campbeltown Loch on the eastern side of the neighbouring Kintyre peninsula became the focus of speculation. A loud splash coming from the direction of the sea attracted the attention of local naturalist John MacCorkindale and the postman Charles Keith. They saw a strange silvery creature with dark streaks around the body and with a giraffe-like head rising out of the water to a height of about 3.7m (12ft), and then splashing back down with the front of its body. MacCorkindale thought he saw a dorsal fin on its back. Had the two observers seen a strange fish? The silvery colour certainly suggests a fish.

The summer following the Loch Scavaig incident (see page 51), another sea serpent sighting occurred at the entrance to Gare Loch off the Firth of Clyde. It, like the Scavaig sighting, was told to Gavin Maxwell by Sandy Campbell. Two fishermen in a coble (a flat-bottomed rowing boat) were taking in lobster creels and heading towards Rhu – a small village to the west of Helensburgh and the centre, in years gone by, of whisky smuggling – when they were stopped in their tracks by the appearance of an object about 10m (30ft) in height. It was 'waving to and fro out of the sea' and was heading at speed towards them. Wisely, they rowed to the shore.

Off Cumbrae Island in the mouth of the Firth of Clyde, a strange sea creature was spotted by a visiting angler on 30 August 1911. In the same area, near Ashton, A H Vincent claims to have seen an animal much like a 'hooded serpent' in the sea a mile south of Clock Light. In August 1953, a camel's head on a giraffe's neck was seen by fishermen in the Firth of Clyde.

The most eerie encounter in this area took place in 1962 at Helensburgh, a seaside resort at the mouth of the River Clyde. Jack Hay and his spaniel Roy were out walking on the beach when the dog started to whimper and cower behind his master. Mr Hay then, according to a report in the *Scottish Daily Mail,* saw something on the beach.

> About 40 yards away I made out a massive bulk with a sort of luminous glow from the street lamps on the Esplanade. It did not move for about a minute, then seemed to bound and slithered into the water. I saw the thing swim out. It had a long body and neck and a head about 3 feet long.
>
> I watched until it was well out in the water and had disappeared. There was a strong, pungent smell in the air.

Mr Hay was scared but, nevertheless, went over to the spot where the creature had slipped into the water. Using a lighted match, he could see a large footprint in the sand. It had three pads and a spur on the back. At about the same time, many people in the beach area of Helensburgh had told of strange noises – 'roaring sounds' – emanating from the beach at the eastern end of the Esplanade. There were also several reports of pet dogs refusing to go out of doors.

South of the Border once again brings us to the north-west coast of England and the location of a most peculiar sea serpent reported by none other than Vice-Admiral Robert Anstruther on board HMS *Caesar* between the Isle of Man and the Irish coast. It happened some time around 1910 and was published in an article by Reginald Pound in *Wide World Magazine* in 1924.

> In the first dog-watch I was standing on the bridge, when suddenly something shot out of the water right in front of me, about half a ship's length off, straight up into the air to about the height of the foremast ahead, about fifty feet.
>
> I, of course, had my galilee-glasses handy, and quickly fixed them on the quadruped – for a four-footed or, at any rate, four-legged, beast it proved to be. In appearance it gave me the impression of a skinned chow-dog, such as one sees hanging up in the butchers' shops of Canton. In shape it reminded me of a chameleon, though a shortened one; the head and short tail also had a chameleon-like appearance.
>
> With outstretched neck and legs it fell, or rather dived, into the sea again.

Anstruther called his navigating officer and, as he reached the bridge, the creature shot clear of the water for a second time.

> It did not appear to have scales, but rather the shiny skin of a reptile. Its feet seemed like the claws one sees represented in figures of Chinese dragons. We waited and waited, but it never rose again.

In 1928, Major W Peer Groves and his family, on holiday on the Isle of Man, chanced upon a sea creature with a head the size of 'a large bull, but rather broader between the ears, ending in a long, dog-like snout'.

The Isle of Man sea serpent. *From a drawing by Michael Peer Groves.*

On the Welsh coast, monster activity has centred on Barmouth, a small holiday town at the mouth of the Mawddach, half-way between Portmadoc and Aberdovey on Cardigan Bay. Sightings have been quite recent.

The first to notice something strange were two people from Colwyn Bay who were walking on the beach near the village of Llanaber, a little to the north of Barmouth, in 1971. They chanced upon some large footprints in the sand at the water's edge. The marks were 30 – 45m (12 – 18in) in diameter. The two walkers thought somebody had been playing a joke and they thought no more of it until 2 March 1975.

It was dusk when six Barmouth schoolgirls, all twelve years old, were walking home along the beach when they disturbed a large unknown creature. It was a little over 183m (200yds) away. The girls were understandably very frightened.

They described it as 'about 10ft long, with a long tail, a long neck and huge green eyes'. They also saw the feet which were 'like huge saucers with three long pointed protruding nails'. They described the skin as 'black, patchy and baggy – it was not like anything we have seen before'.

The terrified girls ran away, but looking back momentarily at the creature in the sea they could still see its eyes above the surface. Next day at school, the girls' art teacher made a sketch from their descriptions, showing a most peculiar-looking beast.

The newspaper stories prompted others to reveal similar experiences. Further to the north, on the north coast of the island of Anglesey, the Minydon Hotel is close to Red Wharf Bay. During February 1975, the employees of the hotel on several occasions saw a strange creature entering the main channel at the western end of the bay. One time, five people watched a black object about 3.7m (12ft) long and protruding about 0.3m (1ft) above the surface swim into the channel. About 9.1m (30ft) behind it a large tail could be seen.

Explanations ranged from basking shark to mini-submarine, although the basking shark most readily fits the description of huge body and tail fin. None of the eyewitnesses, though, mentions a prominent dorsal fin, a characteristic feature of the basking shark.

All went quiet on the Barmouth monster until the summer when there were two sightings from boats. The first was from a fishing boat not far from Bardsey Island, at the end of the Lleyn Peninsula. The experienced fishermen saw a creature with a huge body and a long neck and head surface next to their boat. It appeared three more times during the following hour so they were able to get good views of it, and when they were shown the schoolteacher's sketch of the Barmouth monster the fishermen recognized it immediately.

The second encounter was from a private yacht, a 9.1m (30ft) long sloop, which was sailing about 8km (5mi) from Shell Island (Mochras) near Harlech. On a sunny day and with a calm sea the husband-and-wife crew spotted what they thought was a seal playing with a couple of tyres. As they got nearer they could see that it was no such thing.

> As we drew closer we thought it was a huge turtle, but it turned out to be unlike anything we'd ever seen. It had a free-moving neck, fairly short, rather like a turtle's, and an egg-shaped head about the size of a seal's. Its back had two spines, which were sharply ridged, and it was about 8 feet across and 11 feet long, although the ripples on the water when it dived indicated that it was probably twice that length.

The husband went below to fetch a camera, but by the time he was back on deck the creature was submerging. The couple were shaken and did not sleep for a couple of nights afterwards. Interestingly, although they had been sailing in that area for many years, they had not heard about the Barmouth monster before.

All went quiet again during the autumn of 1975, but in December a man from Dolgellau found large footprints in soft sand near the Penmaenpool toll bridge. He described them as 'a little larger than a good-sized dinner plate', and he saw signs that they were webbed. Strangely, he noted that although the tide had come in and out several times the prints still remained.

Stories of sea serpents around the Welsh coast are not confined, however, to the 1970s. Almost a century earlier a well-reported event took place at Ormes Bay, on the North Wales coast about

30km (19mi) due east of Red Wharf Bay. News of the event reached the auspicious pages of the scientific journal *Nature,* published on 25 January 1883. The correspondence was from F T Mott of Birstall Hill in Leicester.

> Believing it to be desirable that very well-authenticated observation indicating the existence of large sea serpents should be permanently registered, I send you the following particulars.
>
> About three p.m. on Sunday, September 3, 1882, a party of gentlemen and ladies were standing at the northern extremity of Llandudno pier, looking towards the open sea, when an unusual object was observed in the water near to the Little Orme's Head, travelling rapidly westwards towards the Great Orme. It appeared to be just outside the mouth of the bay, and would therefore be about a mile distant from the observers. It was watched for about two minutes, and in that interval it traversed about half the width of the bay, and then suddenly disappeared. The bay is two miles wide, and therefore the object, whatever it was, must have travelled at the rate of thirty miles an hour. It is estimated to have been fully as long as a large steamer, say 200 feet; the rapidity of its motion was particularly remarked as being greater than that of any ordinary vessel. The colour appeared to be black, and the motion either corkscrew-like or snake-like, with vertical undulations. Three of the observers have since made sketches from memory, quite independently, of the impresson left on their minds, and on comparing these sketches, which slightly varied, they have agreed to sanction the accompanying outline as representing as nearly as possible the object which they saw. The party consisted of W Barfoot, J P of Leicester, F J Marlow, solicitor, of Manchester, Mrs Marlow, and several others. They discard the theories of birds or porpoises as not accounting for this particular phenomenon.

The letter prompted further comment and in the edition of 1 February Joseph Sidebotham of Erlesdene, Bowdon, wrote that he had seen the Llandudno phenomenon four or five times before and that he was in no doubt that it was a shoal of porpoises. He went on:

> I never, however, saw the *head* your correspondent gives, but in other respects what I have seen was exactly the same; the motions of porpoises might easily be taken for those of a serpent; once I saw them from the top of the Little Orme, they came very near the base of the rock, and kept the line nearly half across the bay.

Not to be outdone, William Barfoot, J P, from Leicester, responded the following week in the edition of *Nature* published on 8 February 1883.

> Like your correspondent, Mr Sidebotham, I have frequently seen a shoal of porpoises in Llandudno Bay, as well as in other places, and on the occasion referred to by Mr Mott . . . the idea of porpoises was at first started but immediately abandoned. I will venture to suggest that no one has seen a shoal of these creatures travel at the rate of from twenty-five to thirty miles an hour. I have seen whales in the ocean, and large flocks of sea-birds, such as those of the eider duck, skimming its surface; but the strange appearance seen at Llandudno on September 3 was not to be accounted for by porpoises, whales, birds, or breakers, an opinion which was shared by all those present.

And, in the same edition, W Steadman Aldis from the College of Physical Science at Newcastle-upon-Tyne recalled an enormous, fast-swimming monster he saw at Veulettes, on the Côte d'Albatre in northern France.

> Its length was many times that of the largest steamer . . . its velocity equally exceeded that of the swiftest. What seemed to be a head was lifted and lowered, and sometimes appeared to show signs of an open mouth.

Several people were witness to the monster, but what had they seen? A little later the mystery was resolved.

> . . . one day, just as one of these serpent forms was nearly opposite our hotel, it instantaneously turned through a right angle, but instead of going forward in the new direction of its length, proceeded with the same velocity broadside forwards. With the same movement it resolved itself into a flock of birds.

The correspondence, reminiscent of descriptions and interpretations of the events at Skegness (see page 169), continued. This time J Rae of Kensington, London, was reminded of a 'monster' encounter in the Orkneys.

> This was nothing more or less than hundreds of cormorants or 'scarps' flying in continuous line close to the water, the deception being increased by the resemblance of a head caused by several 'scarps' in a cluster heading the column, and by the 'lumpy' seas

> of a swift tideway frequently intervening and hiding for an instant part of the black lines, causing the observer to – not unnaturally – imagine that the portions so hidden had gone under water.

The correspondence ceased, but at the other end of the country, on the Gower Peninsula in South Wales, another sea monster story evolved. The writer and walker A G Thompson recalls in his book *Gower Journey,* published in 1950, standing on the high cliffs overlooking Fall Bay and seeing what he thought was a 9.1m (30ft) long log; that is, until one end moved.

> . . . it became plain that a head like that of a horse with a mane was standing out of the water and watching something on the rocks at the foot of the cliff. What a thrill!! After staring for what seemed minutes the monster appeared to be satisfied and dived with what looked like two distinct undulations of the hind portion. What an uncanny feeling! The writer waited some half-hour or more but there was no reappearance. A local expert was of the opinion that it was a real sea-horse . . .

The Gower is on the northern shore of the Bristol Channel, a funnel of water, with strong tides, tide races and sand bars, that has lured in many whales and dolphins, and the occasional sea serpent.

In 1883, an oily trail, like that sometimes made by seals, was left behind on the surface of the water after a serpent-like marine creature was seen heading out to sea. It was travelling at about 40kph (25mph).

In April 1907, Mr M'Naughton was in a rowing-boat, not far from Portishead on the outskirts of Bristol, when a very strange creature – 'like a huge mummy with sunken eyes enveloped in a sort of hairy flap' – rushed the boat. M'Naughton fell into the water but managed to struggle back into his boat and make for the nearest land, thankful to be alive, but well able to regale the regulars at the local hostelry with his incredible tale.

Further round the coast and we come to the rocky shores and cliffs of the north Devon and Cornwall coasts. They take the full brunt of Atlantic gales and, like the south Cornwall coast, are notorious for shipwrecks, smugglers and sailors' yarns. The steep exposed cliffs, however, are good vantage points to see passing marine life including sea monsters.

The Reverend E Highton saw a sea serpent on 11 October 1882 at Bude in Cornwall. Tintagel Head, a few kilometres to the south, gave two students a privileged view of a long serpent-like monster in 1907. And the same year, Mrs J C Adkins saw a small head on a long neck followed by a series of humps off Padstow.

In 1911, William Cook was taking three ladies from Instow, where the Rivers Taw and Torridge meet, to Westward Ho!, further to the south in Bideford Bay. He changed course to avoid going on some rocks, but the rocks promptly moved – they were alive! The surprised and terrified boatman headed immediately for the shore.

> The thing then stretched itself out in an undulating coil, lashing the water. We calculated its length at from 60 to 90 feet, and its head did not appear. The body was round, and about the size of a 30-gallon cask. Almost black fins, with short intervals, ran the whole length of its back. Its body was of a brownish grey, with scales, very similar to a gigantic snake.

The creature made off to the other side of the Bay, in the direction of Clovelly.

The only other recorded sea serpent sighting occurred in 1906 and was from Land's End. The transatlantic ocean liner *St Andrews* of the American Phoenix Line was passing this, the most south-westerly tip of mainland Britain, when the first and third officers and an American passenger spotted a 5.5m (18ft) chunk of an animal with a body 1.5m (5ft) in circumference and with a set of jaws with 'great fin-like teeth'.

Which brings us back to Cornwall and Madgy Figgin's chair, the Gwavas Lake, the Silver Table, the lovers of Porthangwartha and the Mermaid of Seaton . . . but that's another story.

3

The North American Experience

Thursdays are good for sea monsters!

God created sea monsters on the fourth day of the Creation, so the Old Testament would have us believe, and as a consequence maritime legend has it that the fourth day of the week is the best time to spot them.

Whether or not all sea serpent sightings around the coasts of North America have occurred on Thursdays is not recorded, but there has certainly been no shortage of encounters. New England and Chesapeake Bay have been centres of activity on the Atlantic seaboard, while British Columbia and California are the focus of reports from the Pacific coast.

The earliest observations were made by the indigenous peoples, long before the European colonists arrived. Sea monsters feature in the folklore of most coastal Indian tribes. The Hurons of the St Lawrence Valley, for example, have their *angoub,* and the Chinook of British Columbia are in awe of the *hiachuckaluck.*

Today, the *hiachuckaluck* appears to be alive and well but has gained a new name. It has become known as 'Caddy' the cadbosaurus which, along with 'Chessie' of Chesapeake Bay, the New England sea serpent, and the Stinson Beach Monster of California, has been elevated to national and even worldwide status, recognized by scientists and lay public alike as a phenomenon worthy of further investigation. Indeed, two of the three monsters have been the subject of respectable study – a report from the University of British Columbia on Caddy and the Enigma Project focusing on Chessie.

Chessie

Chessie is a not a new monster. There have been strange creatures reported in the area of Chesapeake Bay since the British were sailing here. But Chessie caught the public imagination in 1977 when it received its nickname in an article in the *Richmond Times Dispatch.*

The Bay is a large tongue of the Atlantic Ocean that, for the main part, slices into Maryland and Virginia on the east coast of the

Doctor Brix

This giant squid was found alive off Bergen by a Norwegian fisherman.

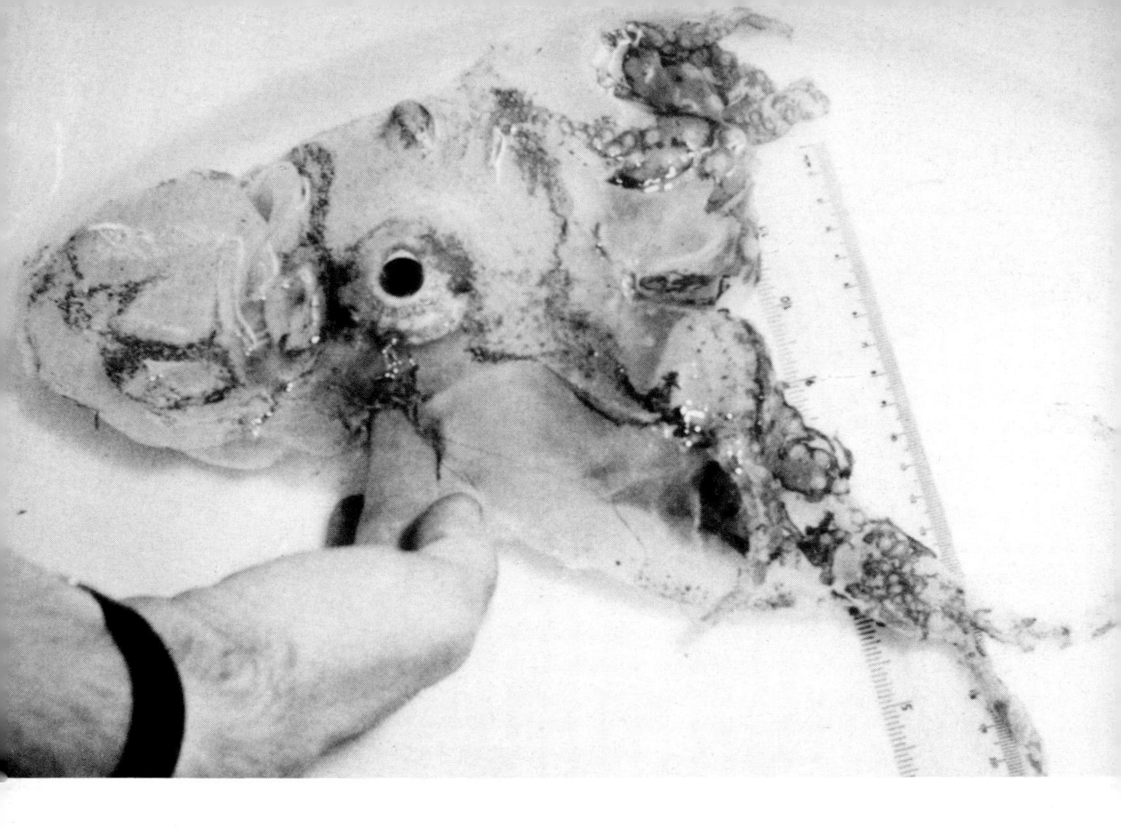

Malcolm Clarke

The jelly-like mass with the big eyes (*above*) is a gelatinous octopus which was washed ashore on the Atlantic coast of Ireland. The giant squid (*below*) came ashore on the Scottish west coast.

Trustees of the National Museums of Scotland

Planet Earth Pictures – Ken Lucas

A sturgeon (*above*) from the Caspian Sea. Note the plate-like scales on the back which, in a large specimen, could account for sightings of creatures with multiple dorsal fins. Groupers (*below*) can grow to immense sizes and are reputed to be able to swallow people.

Planet Earth Pictures – Christian Pétron

BBC Hulton Picture Library

The basking shark carcass caught by a Scottish shark fishery (*above*) can rot in such a way that it resembles a plesiosaur, like the one (*below*) caught by Japanese fishermen off New Zealand.

Taiyo (U.K.) Ltd

USA. Rivers flowing into the inlet include the great Potomac. The largest of the cities nearby are Washington DC and Baltimore.

Observations of Chessie that have reached the Enigma Project have been unusually consistent. Michael Frizzell, the Project Research Director, writes:

> Descriptions of the creature usually denote a 30 to 40 feet long, appendage-free, snake-like body, uniformly dark in color, possessing an elliptical, football-sized head.

Chessie or its ancestors have been seen many times in the past but the earliest record of the present spate of sightings on the files of the Enigma Project is dated 1965. Pam Peters spotted Chessie in the Hillsmere section of South River near Annapolis. In July 1977, Greg Hupka saw it and photographed it at the mouth of the Potomac River. The picture was fuzzy and inconclusive. And there were many more encounters over the next few years.

Favourite sites seem to have been Love Point on Kent Island, the mouth of the Potomac River, and Eastern Bay. The creature has been seen mainly during the summer, between May and September. Is this because more people are on or near the water during the summer months, or does Chessie have an annual migration that takes it into Chesapeake Bay during the summer?

Bill Burton, sports-fisherman and award-winning outdoor editor of the *Baltimore Evening Sun,* who has written extensively about the more recent sea serpent reports, believes there is a link between the arrival of large numbers of bluefin in the area and Chessie sightings. Indeed, in 1984, when there was little bluefin activity early in the season, there were fewer serpent reports. In the autumn, after the bluefin have left, sightings again are few.

Mike Frizzell supports this notion. He collected together all records of Chessie sightings and plotted them on a map of the Bay. They start in April or May near the Potomac River and, as the weather gets warmer, they occur further north. In September and October sightings cease, although it is not clear whether this is because the creature has moved away or because fewer people go boating on the Bay after the autumn.

As for the bluefin connection, there was at least one sighting that was linked directly to bluefin activity. One of Bill Burton's friends remembers being on the flying bridge of his boat, cruising past Caddy Corner where many bluefin were jumping clear of the water. Suddenly, all activity stopped. 'It was very eerie,' recalls the friend (who wishes to remain anonymous). Then, Chessie appeared, ignoring the boat and quietly swimming away.

When I spoke to Bill in 1984 for my BBC radio programme *The Great Sea Monster Mystery,* he knew of at least 78 sightings, many of which had been by trained observers - a coastguard, naval officer, two airline pilots, an ex-CIA official, and an FBI agent. The reports also suggested to him that there may be more than one serpent.

In an article in the *Chesapeake Bay Magazine,* he described the experiences of the Reese family of Mathews, Virginia. They live beside the East River, a tributary of Mobjack Bay above York River. Early one March morning Bill Reese saw a greenish-brown, 6.1m (20ft) long, 'hump-backed thing' swim upriver, round the 17th green channel marker at the mouth of Woodas Creek, and then swim back down again. It moved its 30cm (12in) diameter body with vertical undulations. When he told his wife, she mentioned that she had seen the thing herself the previous year but had been too embarrassed to tell anybody.

In May, Bill Reese was up early again and saw a smaller 3.1m (10ft) long version of the creature near his dock at Lyles Cove.

> I got in my rowboat, grabbed the oars and started after it. It swam to the other side of the cove, would go underwater and then come up just like the bigger one. The head, with eyes like a snake, was held out of the water about as high as a fist from the elbow. Matter of fact if you make a fist and turn your hand away, that is what it looked like.

On another occasion Bill Reese was with his eighteen-year-old son, David, when they saw Chessie swimming out of the cove. In August, their ducks began to disappear!

At four o'clock one morning Bill Reese was woken by the barking of the neighbourhood dogs and a commotion in the marsh grass next to the cove. Next day he discovered his last duck had gone and that was the last they saw of Chessie.

The creature, it seems, is not confined to the water. In 1978, oyster fisherman James Dutton found tracks that, in his estimation, were made by an enormous snake-like animal. Apparently, it had crossed a field and entered Nanjemoy Creek, a tributary of the Potomac River in Charles County. Two fishermen, he claimed, had seen the creature in the creek and were so scared they abandoned their boat.

Dutton also had a rational explanation for the animal. When he had come to live in the Chesapeake Bay area forty years previously, there were many old ships that were left to rot. Some would have traded the South American routes. Could one of them have had a South American reptile, such as the freshwater-living anaconda, on board which subsequently escaped into the waterway, adapted to brackish water, survived the cold winters in warm-water outflows such as sewage outfalls or electrical power station effluents? Certainly Chessie has the size and shape of the legendary giant anacondas beloved of adventure comics, although there would have had to have been either a pregnant female or several specimens introduced to explain the multiple sightings.

1978 was a good year for Chessie. Bob White of the Virginia Beach Coast Guard Auxiliary saw it and reported 'bulging eyes and the skin of a dolphin'. Mary Lewis spotted four or five 'Chessies' near the mouth of the Potomac River on 27 June, and retired CIA employee Donald Kyker saw four on 21 July near Heathsville and, as is the custom in many parts of the USA, he shot one. Four days later, the Smoot family saw three, one large and two small, in the Potomac River near Falls Church. What happened to the corpse of the other one? Nobody knows.

The shooting incident illustrates one of the fears that haunt people like Michael Frizzell and Bill Burton.

'We're constantly afraid,' says Bill Burton, 'that if someone does see this creature and they are armed, like a wildfowl shooter or a commercial fisherman, they may do something to this creature just to find out what it is.'

As a consequence, the organizers of the Enigma Project and others have asked for a resolution to be passed that Chessie should remain unmolested.

In 1979, Eastern Bay was the place where Rosamond Hayes saw the serpent on 13 August. Windmill Point was the encounter site for Gary Speiss on 7 September. Linda Worthington spied Chessie from her condominium at Gunpowder Cove on Rumsey Island on 14 September. It was moving very slowly in about 2.4m (8ft) of water about 46m (50yds) offshore.

In 1980, according to reports in the *Baltimore Evening Sun,* Goodwin Muse, a Westmoreland County farmer, made the first sighting. On 14 June, he and some friends saw a 3.1 – 4.3m (10 – 14ft) long snake undulating in the water at Coles Point, Virginia. In an interview with the Indianapolis *Star,* Muse said:

> I had spyglasses and was up on a bank 18 feet high, and out 30 yards in the river you could see this long, dark streak in the water.
>
> The creature's body was about as big around as a quart jar, and its head was as large as my hand or more.

The six people watched the creature for about fifteen minutes.

About 24km (15mi) to the south, Richmond finance company

manager G F Green III and his family were water-skiing when they saw a 7.6m (25ft) long version. It showed a head occasionally and several humps. The date was 22 June 1980.

During the next few months, according to the files of the Enigma Project, Chessie was spotted: by Doris Peterson at Forked Creek on the Magothy River about 4km (2½mi) west of Chesapeake Bay on 3 July; by David Richardson at the mouth of the Chesapeake off Fort Woll near Fort Munroe, Virginia, in August; by Wayne Henninger at the Wye River Bridge near Eastern Bay, also in August; and by Mrs Awalt at Brackish Pond near the Magothy River again in August.

The next report was in the *Virginia-Pilot.* Coleman and Trudie Guthrie were sailing between Tilghman and Bennett Points at Eastern Bay, Kent Island on 13 September. Trudie Guthrie said they saw 1.8 – 2.4m (6 – 8ft) of a shiny, olive-drab-coloured, swimming object appear about 7.6m (25ft) from their sloop. She thought it was a crab-pot marker until it lifted its head as it slid by.

> This creature had no visible scales, fins, gills, flukes, and no end. It was about 2½ feet thick at the widest spot, and what we could see of what appeared to be the beginning of its tail section was at least 18 inches in diameter.
>
> It made an almost rippleless dive before disappearing. It travelled on a straight line, no slithering, no tail-wagging, no vertical movement, just seemingly unconcerned.

At Smith Point, 129km (80mi) south of Eastern Bay, Chessie had a huge audience. On 28 September 1980, no fewer than four charter boats with about 25 people on board were fishing off Smith Point, Northumberland County.

Bill Jenkins, skipper of the *Miss Cathy II* of Reedville, told the *Times Dispatch* of Richmond, Virginia, what they had seen:

> It was a serpent-like thing. It was swimming with its head out and then it started toward my boat. I've been a charter boat captain for 35 years and I've seen a lot of porpoises and turtles. This was different.
>
> Another boat called my attention to it. We started kidding one another that it was about to eat Don Kuykendall's boat, then it began heading for my boat and we made out that we were going to gaff it, but you couldn't have done that. It was too big.

On board the Kuykendall boat, Don and wife and three others saw the creature. They described a turtle-like head which they estimated to be about 23kg (50lb) in weight. The tail, they said was larger than a human thigh. It and the back were spiked.

Donald Markwith, skipper of the *Midnight Sun,* also saw the creature.

> Its head was larger than a football but not larger than a basketball. It resembled a turtle, but I only saw it momentarily.

During 1981, there were a couple of sightings. Regina Baisley saw Chessie at Swan Creek Bay near Gunpowder River on 8 August 1981. The mouth of the Magothy River, between Weir Point and Miller Island, the Patuxent River, and the mouth of the Chester River also feature in unattributable Chessie reports. A boat-load of fishermen, according to Harry Weishaar of the 'Anglers' fishing tackle shop, were shocked by the appearance of a strange creature just 3.1m (10ft) from their boat moored off Gibson Island. Captain Bill Meadows of the charterboat *El Toro* told journalist Bill Burton about encounters with a snake-like monster while fishing bluefin in the mouth of the Potomac. And commercial fishermen hauling nets off Love Point were scared by a similar beast.

On 24 May 1981, Kathryn Pennington took a colour photograph of a serpent-like creature she saw at 6am, a third of a mile from the shore at Chancellor's Point on the upper Choptank River in Talbot County. She offered the photographs to various learned bodies, including the Maryland Department of Natural Resources, but nobody was interested. That is, until a year later, for the creature seen by Mrs Pennington was identical to that seen by the Frew and Rosier families – whose sighting was to create quite a stir.

In this, perhaps the most exciting encounter, on 31 May 1982, Chessie appeared near Love Point on Kent Island. Businessman Robert Frew, an executive with a laser company, was at home with his family on Memorial Day. They had some friends over for lunch and were in the breakfast area of the kitchen (the back of the house is entirely glass) looking out at the view – 23km (14mi) of Chesapeake Bay right across to Baltimore – when they saw Chessie.

Karen Frew told me that their friend Charlotte Rosier was first to see the creature. At about 7.30pm, she pointed to a dark object about 30m (100ft) off the dock. It was about 2.1m (7ft) long on its first appearance. Robert Frew thought it to be an otter, but when it reappeared for the second time, more than 6.1m (20ft) became visible. On its third appearance at the surface, it seemed even longer and, comparing it to the Frews' 11m (36ft) swimming pool, the assembled eye-witnesses thought it to be over 9.1m (30ft) long. There were humps about every 0.6m (2ft), and the creature moved against the current at about 4 – 5 knots.

Robert Frew watched at first through binoculars but then remembered his video camera. He raced to his bedroom, seized the camera and, from the bedroom window, taped the object for about three minutes. By this time the creature was about 60m (200ft) from the house. Karen Frew described to me what they saw when they replayed the tape.

> It shows two or three head shots – the head and the first seven or eight feet. At one point it came up and the whole thing surfaced and that's when we saw all thirty-five feet of it. One of the most notable things is the absence of any clear-cut markings. It was just very dark in colour, from very dark brown to black.

As Robert Frew taped the event, he noticed that the animal was approaching some children frolicking in the water. Concerned that they might be in danger the Frews and Rosiers yelled to warn them, but they did not hear.

> It was probably the children playing in the water that caused the thing to dive.

Fortunately, the object by-passed the swimmers and did not reappear until further along the bay, whereupon it continued its leisurely progression northwards. The two families raced to their cars and went to the Point in the hope of seeing the animal going up the Chester River, but they never saw it again.

The tape was first analysed by Michael Frizzell and Robert Lazzara, founders of the Enigma Project, and by technicians at Western Electric. They thought it was not a fraud and were able to interest none other than the Smithsonian Institution in Washington. A special meeting of eminent scientists was to be convened.

In the meantime, Chessie continued to appear. Thomas Beery saw it at Bloody Point on Kent Island on 9 July, Lee Smith spotted it at Chesapeake Bay Bridge near Sandy Point on 11 July, Jack and Elizabeth Cundiff witnessed its appearance at the mouth of the Gunpowder River on 12 July, and Clyde Taylor and his daughter Carol chanced upon it on 16 July when walking on the Cloverfield's Community Beach by the mouth of the Chester River not far from Love Point on Kent Island. It was evening, Clyde Taylor told me, and having eaten at a local restaurant they were taking an after-dinner stroll on the beach:

> My daughter noticed something come up onto the surface of

water. The surface at that time was like a mirror, there wasn't a ripple, it looked like a piece of glass. There were no waves, no boats in sight. Its head appeared and Carol said to me, 'Dad, what is that?'

We followed it along the short stretch of beach and the monster showed, as I remember, five humps, and his head showed above the glassy surface of the water. As he came to our right, he was running into a cul-de-sac where a rock jetty protruded out from the sandy beach and he couldn't go any further unless he came on shore.

Meanwhile my daughter was off to his other side, by a bulkhead on the rock jetty and I'm on the opposite side. I'm looking at his right and she's looking at his left.

He raised his head slightly, as I remember, and he noticed the motion of Carol, who was walking towards him. At that, he turned his head slowly, like you turn the head on a ventriloquist's dummy – it almost went around 360°, it seemed to me – and he put his head into an arc. He didn't just sink into the water, to me he put his head down into the water and disappeared, just slid out of view from alongside the rock jetty.

Now, in colour he was black, brown or amber and I say he was thirty feet long, I don't think he was as thick as a telegraph pole, but he was very thick, at least 12 inches. His head had the appearance of a football, except the nose was blunt. I couldn't see any scales or appendages, and I saw the highlight of his eye as he turned his head to look at my daughter. The eye looked like a serpent's eye, like a large snake eye; it had a light yellowish-green tinge to it. It looked quite big, about a couple of inches across.

It moved in [vertical] undulations instead of snake-like, a snake would slither side-by-side and progress in a loop across the ground, but this was like a roller-coaster, up-and-down loop and all loops were the same size as it swam. When it came to a stop, its head remained in a stop position even though its body just drifted in behind it and sank to the shallows where it came to rest. It moved at about 10mph.

It saw me first but I didn't move – I thought it was coming ashore . . . it's my belief that it comes ashore at night to eat, while the water is dead calm and there are no boats in sight.

I could see no markings on the body – it was just a long tube, like an anaconda or python. It didn't look like a fish, but like a giant serpent.

That sums up my sighting of Chessie; I still didn't believe it, I stood there thunderstruck.

Carol Taylor was able to confirm her father's story.

> As I recall, I was standing on the end of the stone jetty and I looked out to my left and I saw a little wake in the water as if a dog was swimming and there was a slight wake behind its nose as it swam. I said to my father, 'What is that?' and he said he didn't know. So we walked back off the jetty and started to walk up the shoreline towards it.
>
> It was, maybe, fifty yards offshore when we first saw it, and as we watched it more and more of the animal appeared – first one hump and then another, and it swam towards the shore and then it levelled out and swam parallel to the shore. There were maybe three or four humps by this time, and then it made another turn and swam towards the shore again.
>
> At this point, I wanted to see it closer so I ran towards it. It never occurred to me that I might be afraid of the animal. And it made another turn towards the shore and I would say there were about five humps up out of the water, all the same height.
>
> As it got, maybe, twenty feet from where the water met the land, and I was maybe ten feet from that, it appeared to sense my movement towards it and, in my view, it just disappeared from sight. It just vanished.
>
> I saw a creature that appeared to me to be a snake, between 25 and 30 feet long, that appeared to be as big around as a telephone pole, that's a dark brown or dark green colour.
>
> It had a head shaped like a rugby ball, but larger than a rugby ball. I did not discern any eye or any mouth, any scales, any appendages or fins of any sort. It just appeared to be a smooth snake-like creature.
>
> There was no way that it could have been someone faking something, there was no one in sight, there were no boats around, the water was only about knee deep – I had a friend of mine walk out into that area the next day to try to measure how deep it was – so there was no way anyone could have been underneath the surface of the water acting like they were an animal.
>
> There was no doubt in my mind that it was a big snake.

The time of the sighting was 8.20pm and the light was going, yet Clyde Taylor and his daughter saw enough to prepare a series of skctchcs. Carol, too, fclt thc animal was coming ashore and, if she hadn't disturbed it, it would have crawled up right in front of them. Carol Taylor told me, though, that she was not frightened by the creature, as there had been no reports of its approaching people or harming anyone in any way, and she would run towards it again if the chance ever arose. But, as she said, 'It was a once in a lifetime thing.'

A month later, on 20 August, the Smithsonian meeting sat in Washington DC to evaluate the contents of the Frew tape, to view Mrs Pennington's picture, and to consider the Taylor sketches.

'Animate, but unidentifiable,' was the consensus of the panel that was led by George Zug, chairman of the Department of Vertebrate Zoology and a member of the board of directors of ISC.

'The usual explanations of partially submerged logs, a string of birds or marine animals, and optical illusion,' Zug stated after the three-hour meeting, 'seem inappropriate for the dark, elongated animal.'

The six other Smithsonian scientists who looked through the material were Joseph Knapp (Director) and Frank Ferrari of the Smithsonian Oceanographic Sorting Center, Nicholas Hutton of the Department of Paleaobiology, James Mead and Charles Potter of the Division of Mammals in the Department of Vertebrate Zoology, and Clyde Roper, Chairman of the Department of Invertebrate Zoology. In a special report to the editor of *The ISC Newsletter* George Zug summarized the findings of the group.

> Throughout the videotape, the object appears [surfaces] and disappears [submerges]. In none of the three major surfacings does much of the object show above water. Also, the object surfaces and submerges in a bobbing or vertical manner rather than a rolling above and below the surface as in most swimming animals. In two of the surfacings, a strong angular structure protrudes from the water, perhaps a foot above the surface; this angular structure is the presumed front end of the object, and was described by one of the witnesses as shaped like a [an American] football. No eyes, ears, or mouth were apparent on this projection when viewed on the videotape.
>
> Simultaneous with the appearance of the 'head', a portion of the 'back' was above the water, but only barely so. The total length of the object above the surface was perhaps three feet, probably no more than five feet . . . The estimates of 10 – 12 feet by the eyewitnesses derive from a dark shadow that they assumed was a rearward continuation of the object beneath the water. The shadow was clearly evident in the videotape, but may have been the rearward extension of a wake. The eyewitnesses reported a vertical undulation of the object. This was not apparent from the tape, nor did it appear that there was a horizontal undulation. The object was producing a low wake, but the manner of propulsion could not be determined.
>
> All of the viewers of the tape came away with a strong impression of an animate object, certainly not some kids

> swimming in a plastic garbage bag . . . We could not identify the object.

Also along for the ride at the Washington meeting were representatives from the Maryland Department of Natural Resources (the people who took no interest in the Pennington photographs), the National Aquarium, various Chesapeake Bay organizations, and a collection of eyewitnesses and investigators.

Craig Phillips, of the Division of Hatcheries and Fishery Management Services of the US Fish and Wildlife Service, and formerly Director of the National Aquarium, was there and he wrote to *The ISC Newsletter* (Winter 1982) with his own conclusions.

> To my eye, this did not appear to be an artefact (floating branch, etc.), but some kind of living creature: elongate or serpentine in shape without visible appendages, or possibly two or more things swimming close together or in tandem, If I recall correctly, its estimated length was 40 feet; I cannot think of any known creature that would exhibit this combination of size and shape.

In his letter, Craig Phillips considered a variety of animals that might be rational explanations for the creature on the tape. He rejected moray or conger eels, for they normally stay below the water. Only a sick eel would be showing itself at the surface. Sea snakes are confined to the tropics, and large constricting snakes such as pythons and anacondas are not usually found in salt water, they do not function well below 70°F, and they would head straight for the shallows anyway.

Phillips, in fact, tried an experiment with a tame anaconda in 1953. He released it several times into the surf at Crandon Park Beach, Key Biscayne, Florida and on each occasion it unerringly swam back to shore.

He similarly rejected an oarfish because the silvery, flattened body in no way resembles the rounded, tube-like body of Chessie. He did not reject totally the concept of fish, muskrats, otters or harbour seals swimming in a line, but could see no reason to consider sharks or dolphins because one would immediately spot the dorsal fins.

The lack of firm conclusions from the Frews' videotape prompted some of those present to suggest that the tape be put through a process of computer enhancement in an attempt to get more detail. This was done, and the 'Chesapeake Bay Phenomenon', as it came to be known at the Johns Hopkins Applied Physics Lab, was suitably enhanced.

Andrew Goldfinger, a physicist specializing in image processing and computer science, and his colleagues, took two years to complete the enhancement, working only in their spare time. They were able to make the object stand out a little more from the background by colour enhancement, but were not able to distinguish more details.

'The film and the photograph [the Pennington picture],' said Mike Frizzell, 'make it apparent this is not a gill creature. It is spending too much time out of the water. The lateral movement indicates it could be a snake.'

Chessie made its 1983 debut in the Potomac River, near Fort Washington, in March. Maxwell Franklin, of Oxon Hill, according to a short report in the *Annapolis Evening Capitol,* was fishing with a companion when they saw an object the size of a telegraph pole. They reported it to the Natural Resources Police and told an incredulous Captain Frank Wood:

> It had the body of a snake, the head of a fish, and the eyes of an alligator. It cut through the water like a boat.

The creature swam to within 2.4m (8ft) of where they were fishing, and when they shone a torchlight at its head, its eyes 'glowed green'.

The next sighting was at 3 o'clock in the morning on 27 April. The attention of an Alexandria charter boat owner (who again wishes to remain anonymous for fear of ridicule and loss of business) was drawn to an object in the water by his son and a friend. He put the spotlight on it and saw a 9.1m (30ft) long, smooth-bodied creature glistening in the beam. As it passed within about 30m (100ft) of the shore alongside the Prince Street Dock, they could see, illuminated by the lights cast by night workmen repairing the Woodrow Wilson Bridge, its head bobbing up and down in the water. The workmen failed to notice the object and when the boat owner reported the event to the District of Columbia Harbor Police it was written off as a pile of 'trash'.

During the summer, boatman Ernest Bilhuber encountered Chessie in the South River above Riva Bridge, and St Michael's waterman David Jump and his brother Jesse saw it while crabbing in the northern Bay area.

On 1 October, John Hoffman, a 51-year-old Baltimore businessman who has fished in Chesapeake Bay since he was eight, and Ruben Ribaya were entering Herring Bay in southern Anne Arundel County, when they saw a 7.6m (25ft) long creature which was 'greenish, dark brown in color.'

'The wake,' said Hoffman, 'was coming out of it like someone waterskiing.'

News American staff reporting the event approached the National Marine Fishery Service in Washington for their comments but were fobbed off with porpoises, a wayward whale or a school of cow-nosed stingrays for an explanation. True, sharks, rays, dolphins and whales do enter the Bay and, indeed, a whale became stranded and died near the Bay Bridge in 1975, but Hoffman and Ribaya were not convinced the creature that passed so close to their boat could be explained away so easily.

Ken Welder would agree. His encounter came on 29 August 1984. He told Bill Burton how seven (four adults and three children) of the nine people (two young children were below) aboard his 25-foot powerboat were watching cow-nosed rays coming to the surface of a calm sea. It was about 6.45 in the evening and they were 4.8km (3mi) south-east of Bloody Point Light on Kent Island. Suddenly, the rounded head of a strange sea creature appeared just 15.2m (50ft) from the boat. It was about 20cm (8in) in diameter and came about 46cm (18in) out off the water. Welder's wife Gail screamed and it looked towards the boat, but did not seem to be afraid or aggressive. It opened its mouth, which was like that of a turtle, although they could see no shell. Ken Welder, who has seen turtles and seals before, thought it did not look like a sea turtle, anyway. It did, however, 'gasp' as if breathing, according to Ken. His wife said it 'gurgled'.

The following year, a Senate Committee on Economic and Environmental Affairs sat in Annapolis. On the agenda was a resolution tabled by Senator George Della Jr asking the state to encourage more scientific inquiry into Chessie's possible existence, and also consider protecting it in case the animal were ever caught. An article in the *Baltimore Evening Sun* by Dan Fesperman reported that the committee's chairman, Senator Norman Stone Jr, said, 'When we find out what it is, we may not really want to protect it. We may want to get rid of it.'

Mike Frizzell tried to convince the committee otherwise but met with a brick wall. The Maryland Department of Natural Resources relented a little and declared it neither supported nor opposed the resolution but it did go so far as to say they would protect 'he or she or it or whatever else'.

On 5 February, the resolution was finally defeated and the committee would not even recognize the 'possible existence' of Chessie.

'I can't see it, so I can't vote for it,' said Senator Arthur Dorman.

On Good Friday 1985, Chessie tried her/his/its best to oblige the good Senator. Nancy Gabriszeski was having a cup of coffee on her sun porch which overlooks the water off Riverview Road near the north side of the mouth of Back River, when she thought she saw a floating piling. She was totally mesmerized when the piling turned into something that she said 'looked like a snake blown up 50,000 times'. She told Bill Burton that despite having a 35mm camera near by, she was so taken by the creature she didn't think of using it.

The serpent swam around a fixed piling and she could see 'camel-like humps' on the back of the greenish-brown object which turned out to be arches of the body. It slithered into about 2 – 2.5m (6 – 8ft) of water, with its head well above the surface. It was heading north-east towards Hart and Miller Islands.

She thought the body to be twice as round as a telegraph pole with the head about the same diameter as the body. She could not distinguish eyes but thought she could see indentations where the eyes should have been.

One of the most recent Chessie sightings of which I have information was featured in *The Richmond News Leader* of 24 July 1986. Dentist Jack Bishop saw what looked like a gigantic snake in the Tred Avon River just south-west of Richmond, Virginia. The date was 25 May, and Dr Bishop was standing on the boat dock of a friend, furniture store owner Ken Boundrie, admiring a new boat. He told me about his encounter. The time was 7 o'clock in the evening and the sun was setting. The creature appeared about 91m (100yds) away, in the middle of the channel, heading west towards the mouth of the Choptank River.

> I saw, coming out of the water, the head – the front-end of it – and it was snake-like, it was pointed. I couldn't see any eyes because it was too far away. As it came up out of the water, I could see the sections behind come up at the same time; I could see three sections and it undulated, that's how it swims – up-and-out, up-and-out, over and over again, and it swam on out like that, going about 5 – 6 knots; it was moving along pretty good. It was a reddish-brown colour.

They estimated the length to be about 6.1m (20ft) and the diameter about 46cm (1½ft). They watched the head and the arches going up and down for about two minutes. After the newspapers had reported Jack Bishop's sighting, many of his patients 'came out of the closet' and admitted to having seen the creature themselves, and some made significant observations.

People who have seen this thing up close have told me that they

> have followed it along in the water for 45 minutes and they saw it had snake-like markings and it had some texture to the skin, it was not perfectly smooth. When they saw it, it was not swimming up out of the water but just below the surface, just taking its time.

The people involved were Jan and Bob Snead of Easton. Their encounter had been in 1984 on the first cruise of the early spring season. They were going along the Choptank River at 'trolling speed', Jan Snead told *The Washington Times Magazine.* Then, she looked down into the water.

> There was this thing beside the boat – it reached the whole length of it, which is 34 feet. And it was thick enough to put your arms around – like two feet, and just under the surface of the water. I couldn't believe what I was seeing.
>
> It would go below, then resurface, very graceful, with the wake of a small boat – I got the impresson it was as curious about us as we were of it. But it wouldn't allow us to approach it – it would come to us instead. It seemed shy, truthfully, and this went on for about 45 minutes. I'd give anything in the world to see it again – every time we go out now, we watch, we wait.
>
> But there is something strange. Bob and I – we're friends with Jack Bishop, and the day before he saw it this year – we were out in the smaller boat, right in the same area. Well, I got the strangest feeling that the thing – the thing we'd seen two years ago – was there again. I had this feeling. I looked and looked for it. But no luck.

The New England Sea Serpent

New England sea serpents became fashionable in the nineteenth century. There were, suggests Bernard Heuvelmans, a whole family of the creatures making frequent visits to the New England coast to the extent that if they failed to turn up for one season they were sadly missed. Heuvelmans has recorded 130 sightings before 1900, mostly in Maine's Penobscot Bay and neighbouring Broad Bay, and off Massachusetts.

Like Chessie, the New England accounts were consistent. The animal was 15 – 18m (50 – 60ft) long, black, and snake-like, and it raised its head above water to a height of about 2m (6ft). Most times, just one was seen, but on several occasions two have been spotted together. They move, it is said, in the traditional sea serpent manner, with a vertical flexing of the body.

John Josselyn was the first to enter into print. He wrote in *An*

Account of Two Voyages to New England, published in 1674, about a sea serpent that was coiled up on a rock at Cape Ann, at the north end of Massachusetts Bay. Some Englishmen and a couple of Indians were passing in a boat, and the colonists, true to form, wanted to shoot it. The Indians said no. If the creature was wounded it could be very dangerous.

At the end of the eighteenth century, Fox Island was the focus of sea serpent sightings. Many a musket had a sea serpent in its sights, but the year when things really began to warm up was 1817.

On 6 August two women and several fishermen saw a creature enter the harbour at Cape Ann. By the 10th, it had arrived at Ten Pound Island and on the 13th and 14th, thirty or so people, including Gloucester's Justice of the Peace, saw it in Gloucester Harbor. On the 14th, four armed boats were dispatched and one seaman shot at the monster and hit it in the head.

The Linnean Society of New England set up a committee of inquiry consisting of a judge, a physician, and a naturalist. In the meantime, unaffected by its head wound, the creature appeared again on the 15th, and was seen by a whole group of people, including the crew of a ship in harbour. The following day, the crew of a rowboat came so close as to touch it with their oars, and the day after that it was encountered by two men in a sailing boat.

On 22 August, two independent witnesses saw the creature partly hauled out on a sandy beach. Then, sightings began to take place from ships further out to sea – 2 miles east of Cape Ann, for example, on the 28th. It did not reappear again until 3rd and 5th October, by which time it had moved, if indeed it was the same beast, into Long Island Sound and close to New York.

The Linnean Society drew up a list of questions to ask the witnesses and, before the Justice of the Peace, they were asked to submit their affidavits. Oudemans picks out one declaration which I quote here. It was from the man who shot at the beast in Gloucester Harbor.

> I, Matthew Gaffney, of Gloucester, in the County of Essex, Ship carpenter, depose and say: That on the fourteenth day of August, A.D. 1817, between the hours of four and five o'clock in the afternoon, I saw a strange marine animal, resembling a serpent, in the harbour in said Gloucester. I was in a boat, and was within thirty feet of him. His head appeared full as large as a four-gallon keg, his body as large as a barrel, and his length that I saw, I should judge forty feet, at least. The top of his head was of a dark colour, and the under part of his head appeared nearly white, as did also several feet of his belly, that I saw. I supposed and do

> believe that the whole of his belly was nearly white. I fired at him, when he was nearest to me. I had a good gun, and took aim. I aimed at his head, and think I must have hit him. He turned towards us immediately after I had fired, and I thought he was coming at us; but he sank down and went directly under our boat, and made his appearance at about one hundred yards from where he sank. He did not turn down like a fish, but appeared to settle directly down, like a rock. My gun carries a ball of eighteen to the pound; and I suppose there is no person in town more accustomed to shooting than I am. I have seen the animal at several other times, but never had so good a view of him, as on this day. His motion was vertical like a caterpillar.

All the eyewitnesses came to some measure of agreement, with only minor differences of fact. They saw an enormous, dark, smooth-skinned, snake-like animal, that moved very rapidly with a vertical flexing of its body. The Society concluded that the sea serpent was, indeed, a large snake, and therefore, like a sea turtle, should come ashore to lay eggs. The hunt was on for the egg-laying site.

In the autumn of 1817, two boys were playing beside Loblolly Cove, near Cape Ann, and they found a small, one metre (3ft) long, black snake. On the snake's back were little bumps, just like the sea serpent's – or so everyone thought. The boy's father spiked the snake and sold it to a dealer who, in turn, offered it for examination to the Linnean Society. They accepted it, at face value, as a young sea serpent and after dissecting it declared it to be the Atlantic Humped snake with the scientific name *Scoliophis atlanticus*. They said that it was closely related to the Common Black Snake *Coluber constrictor*. Later examination by other naturalists revealed that it was not only closely related to the common black snake, it *was* a black snake, but with deformities.

Scorn was thrown at the Linnean Society, but the fact did remain that the great sea serpent was a reality (unless most of the population of Gloucester were mass-hallucinating). It returned to Gloucester Harbor again in July 1818 and each summer for the next six years. It failed to make an appearance in the summer of 1825, but returned in 1826 and continued to turn up for many years afterwards.

In more recent years the New England sea serpent has not been seen again. There was a flurry of interest a little to the north with four separate sea monster sightings in July and August 1976 off Cape Sable Island, Nova Scotia. Otherwise, attention has been focused further south in Chesapeake Bay. Could the New England Sea Serpent have moved south, encouraged, perhaps, by some environmental or climatic change?

Caddy

The waters along the Pacific coast near Vancouver on the US-Canada border are rich in marine life. The easterly-flowing Pacific Current heads towards the Northern American mainland and splits in two, half going to the south to form the Californian Current and half to the north to form the gyre of the Alaskan Current. With it swim the Pacific salmon that return in huge numbers to the rivers and streams of British Columbia. And following the shoals are seals and sea-lions, which also eat sculpin, greenfish and rockfish, and 'resident' pods of killer whales that dine on the returning fish. Also attending, if reports are to be believed, are sea monsters – three varieties. There is the big-eyed, long-necked version with a huge body that lies hidden just below the water surface. Then there is the small-eyed, long necked, big-bodied one. And third, there is the true snake-like animal – the great sea serpent.

The three types were determined as a result of a survey of sea monster sightings made by Paul LeBlonde and John Sibert, from the Institute of Oceanography at the University of British Columbia. The manuscript was published in June 1973 and has the title 'Observations of Large Unidentified Marine Animals in British Columbia and Adjacent Waters'.

The authors sought information on sea monsters by lecturing, appearing on radio and TV, and writing to marinas, lighthouse-keepers, fishing clubs and local newspapers. The response was good, and they were able to divide the information into actual sightings, second-hand reports, and ancillary evidence.

Their earliest eyewitness account was from Philip Welch, of Port Alberni. His encounter was in 1905 or 1906 while working as a timber feller at Cracroft Island. One Sunday in September, he and a friend took a 16-foot rowboat to the mouth of the Adams River, in thc Johnstone Strait, to go trout fishing. There was, though, an enormous salmon run at the river mouth making trout fishing impossible. At 9am, they were about to give up when a long brown neck appeared about 183m (200yds) from the stern of the boat. They contemplated shooting at it but, deciding discretion was the better part of valour, rowed as fast as they could for the shore. The salmon were everywhere; each stroke of the oar made contact with fish. The monster came closer, moving faster than they could row, and then it submerged, not to be seen again.

Giving more detail about the head and neck, Welch estimated that it was about 2 – 2.5m (6 – 8ft) long, and tapered from a 51cm (20in) diameter base to about 20-25cm (8 – 10in) at the head end.

The head was like that of a giraffe complete with two small 13cm (5in) bumps, and nostrils.

The second sighting was in May 1922 near Pulteney Point lighthouse on Malcolm Island. J Philips and C G Cook were anchored there. The sea was rough and there was quite a wind from the north-west. The two men were on 'stand-by', waiting for the lighthouse tender to arrive. Mr Cook thought he saw the boat coming without its sail up, but what they thought was a mast turned out to be the head and 2.1m (7ft) of neck of a 7.6m (25ft) long strange sea creature. The head, which moved like a snake, had nostrils, large cow-like eyes, and the brown coloured body appeared to be scaly. They breathed a sigh of relief when it passed them by, although they did say they were impressed with how gentle the creature appeared to be.

The Malcom Island sea serpent of 1922. *From a drawing by C G Cook.*

The third encounter was in late December 1937 or early January 1938 on the Oregon coast of the USA, south of Cape Perpetua, about 3km (2mi) south of Yachats and 16km (10 – 12mi) north of the famous sea-lion caves. It was a stormy day, and the two observers, who wish to remain anonymous because they have been ridiculed so much in the past, were watching the enormous waves breaking on the shore in a rocky chasm known as the Devil's Churn. Then, heading towards the Churn, they saw a creature which came to within 30m (100ft) of them. It had a long 4.6 – 6.1m (15 – 20ft) neck with a mane the colour of seaweed all the way down to the body. The head looked, to one witness, like that of a horse or camel, on which two small ears 'fluttered incessantly'. The other thought it was more a giraffe-like head with small horns.

They estimated the body to about 2m (6ft) across with a ridge running along the back. When the creature dropped into a wave trough, one witness saw a long tail which he thought gave the animal an overall length of 16.8m (55ft). In the report, they described what happened:

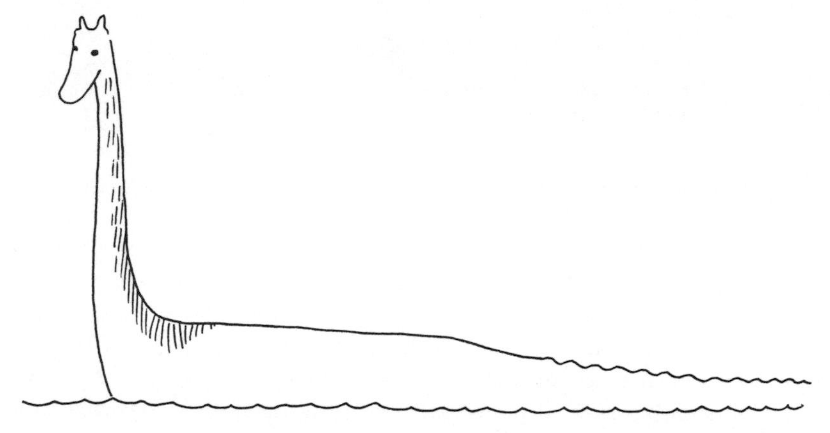

The Devil's Churn sea serpent of 1937.

> It came directly towards the Churn, swimming slowly without any visible propelling motions, and stopped close to the entrance of the chasm for 15 to 20 seconds, during which time it was only about 100 feet from the observers. They remarked that 'the heavy breakers did not seem to toss it around one bit'. A truck came by on the highway during that time; the animal turned its head to look at it, then looked at the witnesses and again at the truck. It then took off southwards along the coast, moving quite rapidly at a speed estimated to be 25 to 30 knots. Both witnesses followed it along the shore-highway in their car. At an observation point about a mile south of the Churn, the animal was about 300ft offshore. It then veered off from the coast at an angle of about 30°. An unidentified man stopped his car at the lookout and they all watched it swim out to sea until it vanished.

In the footnotes, the authors of the report point out that, like the Welch 1905/1906 unidentified animal, this one also had horns.

Continuing the chronological order, the next report is from Mrs M Tildesley and the event occurred one summer afternoon in 1939. She was accompanied by a friend, Mr Duncalfe, and they were sketching the Olympic Mountains to the south from a small beach at Ten Mile Point, near Victoria on Vancouver Island. They stopped for lunch at about half-past one, when Mr Duncalfe glanced up just in time to see a greenish-yellow 'log' whizzing towards the Pacific. It dived without disturbing the water, came up further out to sea, and disappeared round a headland in the direction of Cadboro Bay. In the brief period it was in view, the two observers noted a neck about 2m (6ft) long with a head very much the same width as the neck.

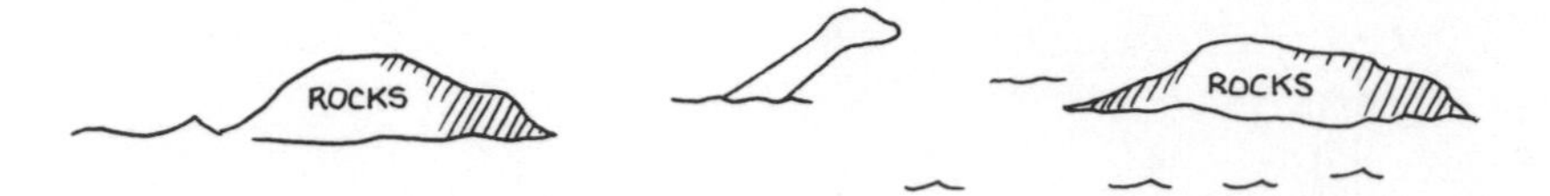

The Ten Mile Point sea serpent of 1939. *From a drawing by Mr Duncalfe.*

That same summer, Captain Paul Sowerby was on the bridge of his ship about 19km (12mi) off Destruction Island on the Washington coast when he saw what he at first thought was a polar bear floating in the water. Thinking this was not at all likely, he manoeuvred so as to get a closer look and stopped engines. He saw a 1m (3ft) long section of head and neck (he thought there was more neck under the water) with huge blinking eyes, and a hide that was ringed with what appeared to be rolls of fat. Captain Sowerby was within about 9.1m (30ft) of it when it submerged. It reappeared astern. The boat was turned about but the crew were looking directly into the sun and lost sight of the creature.

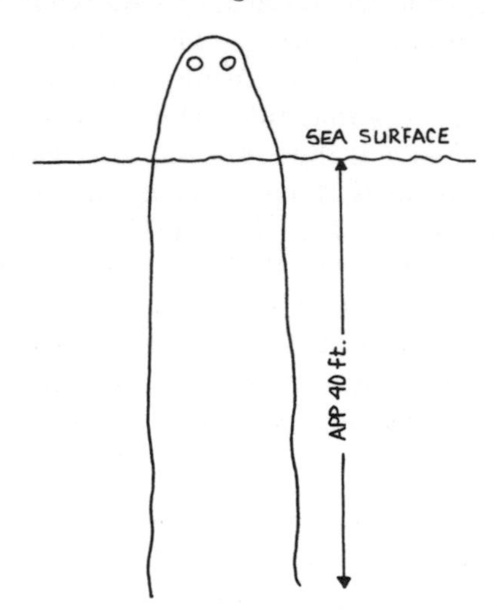

The Destruction Island sea serpent of 1939. *From a drawing by Captain Paul Sowerby.*

The following year, attention was focused on Discovery Passage near the mouth of Rock Bay. Isaac Krook and E H Luoma were motoring northwards through the Passage in a 14-foot boat when, at about half-past two in the afternoon, a rounded, brown-coloured, serpent-like creature appeared about 61m (200ft) ahead of them. They slowed down immediately. The animal then crash-dived, causing quite a commotion as the body behind the neck went into large coils. It left a trail of bubbles in the water and reappeared at the surface about 183m (600ft) astern. It took one look at the astonished men, breathed in, turned, and dived again.

Mr Luoma estimated the creature to be about 15.2 – 22.9m (50 – 75ft) in length, and it had the girth of an oil drum. The largest loop was about 2 – 2.5m (6 – 8ft) in diameter. The loops either side were progressively smaller. The animal had eyes on the side and a centrally positioned mouth, like that of a snake.

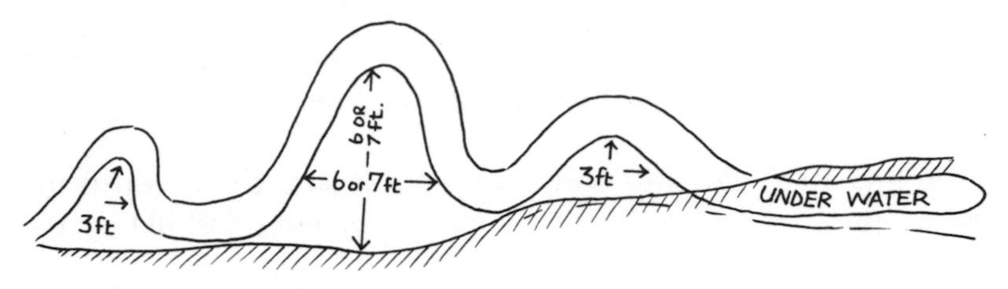

The Discovery Passage sea serpent of 1940. *From a drawing by E H Luoma.*

This animal, as the authors comment, is not like the previous sightings, but is remarkably similar to a 1946 report coming up next and, more important, very similar to three sightings from California in more recent years (see pp 97).

The 1946 affair occurred at about half-past eight on an autumn morning. Mrs Dorothy Nilsen and her husband were in their boat on passage between Southey Bay on Saltspring Island and Chemainus on Vancouver Island. Just outside the bay, Mrs Nilsen saw gulls feeding on herring, but was distracted by the sudden appearance of first one and then two large 1.5m (5ft) diameter loops in front of the boat. They looked like round pipes, the colour of putty, and with a fin along their length. The loops moved rapidly towards the fish, scattering the birds, and leaving a considerable wash behind.

The Southey Bay sea serpent of 1946. *From a drawing by Dorothy Nilsen.*

At the same time, the Nilsens' neighbour, Frank Hayward, arrived. He had also spotted the birds taking the fish and was putting down fishing lines. Later, he told Mrs Nilsen that he too had seen the strange creature. It had surfaced next to his boat and gave him quite a shock. It had a head like that of a sheep, and its eyes, he said were 'wicked'.

In two reports not quoted by LeBlonde and Sibert but referred to by Heuvelmans, there are descriptions of a similar creature. The first was on 10 August 1932 when F W Kemp and his family watched the animal appear, moving against the current, between Chatham and Strong Tide Islands in the monster-rich Strait of Juan de Fuca. It came to the shore, put its head out of the water and up on to the rocks, waving it from side to side. In a statement in the *Victoria Daily Times,* Kemp described what happened next:

> Then fold after fold of its body came to the surface. Towards the tail it appeared serrated with something moving flail-like at the extreme end.

He estimated the animal to be about 24.4m (80ft) long and 1.5m (5ft) thick. The colour was bluish-green, and the skin 'shone in the sun like aluminium'. They saw that the head had a mane which looked very much like kelp drifting round the body.

The Chatham Island sea serpent of 1932. *From a drawing by F W Kemp.*

A year later, on 1 October 1933, Major W H Langley, a barrister and Clerk to the British Columbia Legislature, was sailing with his wife in their yacht *Dorothy* near Discovery and Chatham Islands. The time was about half-past one in the afternoon and they were making good headway when they heard a loud noise which Langley describes as 'a grunt and a snort accompanied by a huge hiss'. Looking ahead of the boat they saw a large unidentified olive-green-coloured creature near the edge of the kelp beds. It had serrated markings along the top and sides. Comparing notes with Mr Kemp, the Langleys felt they had seen a similar, if not the same, animal.

One clear and sunny evening in late August 1943, Jane Easson was on a small 2.4m (8ft) high piece of cliff at Roberts Bay near Sidney, British Columbia. In front of her, the sea was calm. She was feeding gulls. It was about seven in the evening. Suddenly she noticed, just a few feet in front of her, several spines sticking above the sea surface. Below the water she could make out the shape of a 2.4m (8ft) long animal which was causing a slight wake as it moved slowly along. It was not, she felt, a dolphin or a seal.

Another 'spiny body' sighting occurred in 1951. George and Laura McLean and Laura's brother Major W J Wilby were going to the store at Heriot Bay on Quadra Island, fishing as they went, and had reached a place called Surge Narrows. They noticed that all the fish had disappeared and then about 9.1m (30ft) from the boat they spotted a large greenish-grey animal at the surface. It was about 12.2m (40ft) long and had a 0.3m (1ft) tall spiny fin running the length of its back. They could see no head or tail.

It moved at about 8kph (5mph) and undulated up and down. Mrs McLean described the movement like that of a snake, although she acknowledged that snakes move laterally. She did, though, say that she saw it again on two further occasions.

Another sighting picked up by LeBlonde and Sibert was on 23 June 1946. Winnifred Grist was walking her dog on Balmoral Beach in Comox. It was a fine, clear day. The sea was calm. In front of her, about 10.7m (35ft) offshore, there appeared a sea creature with a horse's head. It was watching Mrs Grist's dog. It was seaweed-brown and had red eyes (another feature shared with the Stinson Beach monster. See page 97).

Mrs Grist had her camera with her but as she raised it to her eye the animal submerged and all she got was a picture of the beach!

Nineteen-forty-seven is the year of the next sighting, off Siwash Roch, in English Bay near Vancouver. It was February or March when the Pantages and their friend Chris Altman were fishing from their boat. The creature they saw had a horse's face, two horns or ears, several humps and 'great big green eyes'. It snorted and blew water out of its nose and it swam with the up-and-down motion of a caterpillar.

Rod Kline's sightings in 1948 took place at about noon on successive weekends in April or May. He was seal-hunting near China Point in the Juan de Fuca Strait at the southern end of Vancouver Island and he saw the creature on two occasions. During the second sighting, he shot at it and it rolled over, not to be seen again.

The animal was about 9.1m (30ft) long and had a thick neck. Kline estimated it to be about 3m (10ft) in diameter at the shoulder. It had two small flippers – proportionally smaller than those of a seal – a large mouth and small eyes. The colour was mottled grey, and lighter on the belly than the back. It swam with a bobbing motion at the same speed as a fishing boat, about 7 – 8 knots.

In 1950, Evelyn Leighton was fishing off Piper's Lagoon near

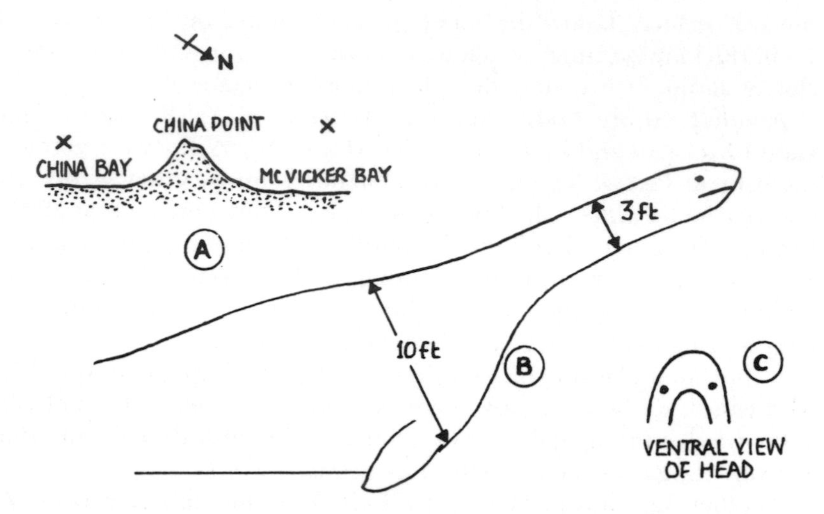

The China Point sea serpent of 1948. *From sketches by Rod Kline.*

Nanaimo when she saw a 'horse's head minus ears'. She shouted to her husband to come and see but her call frightened the beast and it disappeared.

In 1955, Mrs E F Spence was on the Clover Point side of Ross Bay near Victoria when she saw a 6.1 x 0.3m (20 x 3ft) 'log'. Suddenly, a 'branch' rose into the air, as if the 'log' had rotated, and the log itself moved sideways. Closer inspection revealed the 'log' to be an enormous animate body and the 'branch' a 3.7m (12ft) long neck tapering to a snake-like head. It raised its head, looked around, and then placed it back into the water where it waved it from side to side like a waterbird feeding.

On 7 July 1957, N Erickson was in his boat 1.6km (1mi) west of Merry Island lighthouse. He was negotiating the Welcome Pass on a calm, clear evening, just as the sun was setting, when he spotted what he thought was a piece of driftwood with a duck struggling to get on at one end. As he got nearer he discovered that it was a horse-like creature, and that the splashing was caused by it flapping its ears.

Since 1958, Alison McCord's friends have greeted her monster story with 'polite disbelief'. Nevertheless she told her tale to LeBlonde and Sibert. It took place near the seawall in Victoria, at the southern end of Vancouver Island, while Mrs McCord was walking her dog along the path in Dallas Road. The early morning silence was broken suddenly with a sound similar to the 'bark of a dog'. She looked around but could see no dog apart from her own, but then she spotted a pony-sized head in the kelp just off the point.

It was about 76.2m (250ft) away from where she was standing. The head was dark brown and it had a Chinese-type moustache. Behind the head were three humps. The creature barked again twice, but as Mrs McCord drew closer it went straight down in true monster fashion – no dive, just straight down.

The authors draw attention to the bark (the only other time sounds have been heard was in South Devon, see page 33), and the moustache. The moustache, they feel, could be kelp dangling from the snout, but this is not the first time a Chinese-style moustache has been mentioned in sightings – Steller saw one off the Aleutians (see page 112).

A professional fisherman from Victoria, David Miller, had an intriguing encounter with a strange sea creature about 800m (½ mi) south-east of the Discovery Island light in 1959. In the University of British Columbia report, the authors quote the entire letter; this is part of what Mr Miller wrote:

> While engaged in commercial fishing one afternoon in late November 1959, my partner Alfred Webb and I observed this strange creature surface roughly 80 feet on our port beam. It started to move rapidly away from us so we speeded the engine up and gave chase. We got within 30 feet when it suddenly submerged, not in the method seals or sea-lions do but as though something pulled it under. A few minutes later we arrived at its place of submergence and there was a tremendous turbulence

The Welcome Pass sea serpent of 1957. *From a drawing by N Erickson.*

> suggesting a creature the size of a 30-foot sei whale. Its speed under water was also astounding as it surfaced a few minutes later over a hundred yards away. It stayed up while we took off after it again but this time it never let us get close again.
>
> The first encounter was so close that both of us remarked about its large red eyes and short ears visible at that range.

Both fishermen are experienced seafarers, with many sightings of seals and sea-lions. They have even seen elephant seals at sea, a rare experience in itself, but they had never seen anything like this creature. They also pointed out that many fishermen in the area had seen similar beasts but were very reluctant to talk about them. David Miller did mention, however, that his brother had seen one in near the Rae Rocks in the Juan de Fuca Strait just a few months later. Their own creature disappeared eventually in the direction of the Haro Strait.

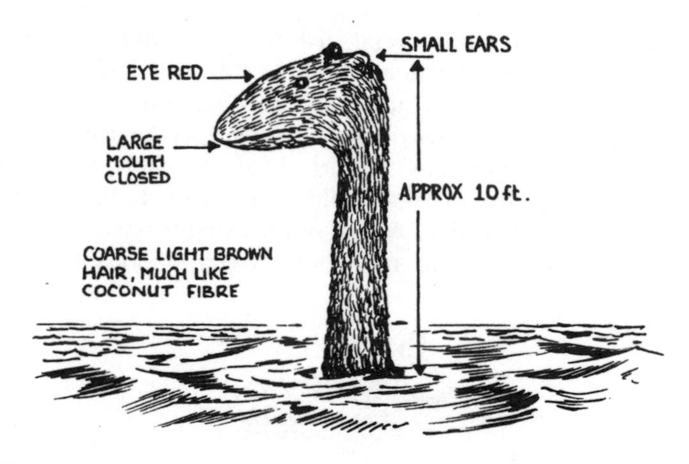

The Discovery Island sea serpent of 1959. *From a drawing by David Miller.*

The Dungeness sand-spit proved to be a suitable vantage point to watch sea monsters for Mrs E Stout and Mrs Parson and the Stouts' small children, aged four and five, in 1961. For one child, the experience was most disturbing. He grabbed his mother and began to cry with fear. The creature that caused this consternation first rose out of the sea about 400m (¼ mi) away, disappeared, and appeared again much closer. Mrs Stout wrote:

> We could see that it was some kind of creature and distinctly saw that the large flattish head was turned away from us and toward the ship (the ship was heading in the direction of Pt. Townsend). I

think all of us gasped and pointed. We could distinctly see three 'humps' behind the long neck. The animal was proceeding westward at an angle toward us. It sank abruptly again and reappeared closer, almost due north of us. In the dim light we could distinctly make out colour and pattern, a long floppy 'mane', and the shape of the head . . . We estimated that the length of the neck was at least 6ft, and the head probably at least twenty inches long . . . I simply could not accept the long floppy mane or fin. Yet we all saw it. It was a limply hanging thing. We deduced that the humps were at least five feet . . . The animal was a rich deep brown with large reticulations of a bright, burnt orange. The pattern wasn't unlike that of a giraffe except much larger. The fin appeared to be drab or colourless.

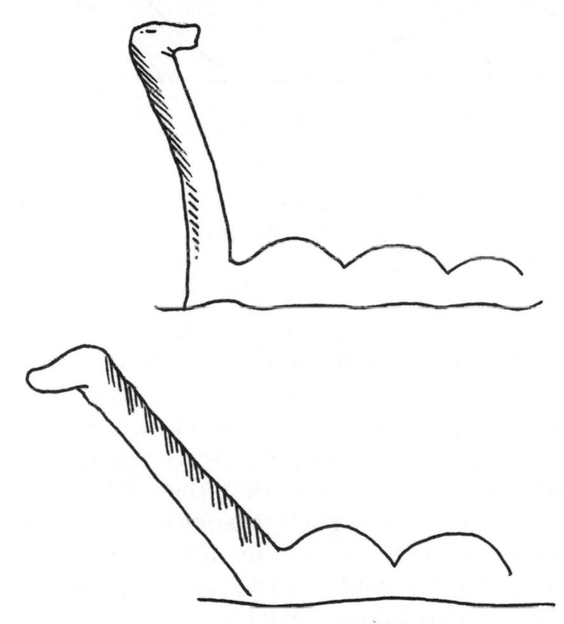

The Dungeness Spit sea serpent of 1961. *From a drawing by Mrs E Stout.*

We could see no body movements, except that the neck lowered and moved backward with grace and the head swivelled, raised and lowered. Its forward progression was smooth, like a swan's. It sank and rose almost perpendicularly; although there was no indication of effort, it progressed westward, toward Pt. Angeles at a fair rate of speed. The whole episode lasting about eight minutes.

Mrs Parson had a camera with her but unfortunately considered the light too dim to take a photograph – another opportunity to obtain more substantial evidence lost.

On 13 February 1968, Jean Scott was sitting on her porch on Crane Island. It was about half-past six in the evening that she heard a blowing sound that she took to be a whale. On looking out to sea, she saw, not a whale, but a strange-looking beast, blowing from its mouth rather than from a blowhole. The animal was about 45.7m (150ft) from the shore. It was 'very large', with 'grey skin', and 'lengthwise ridges'. It had a long neck and a 46.61cm (18 – 24in) long head like that of a cod.

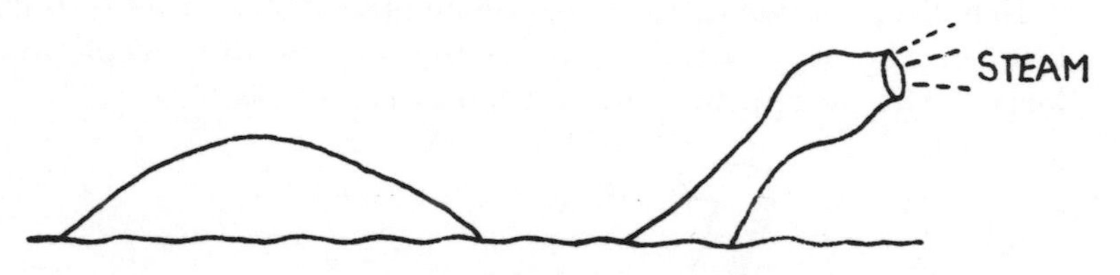

The Crane Island sea serpent of 1968. *From a drawing by Jean Scott.*

Mrs Scott and her husband watched it until sunset. It kept putting its head below, as if feeding on the bottom, and raising it every so often to blow. It moved slowly, except when a gull came to sit on it and it wisely moved rapidly away. Mrs Scott cannot work out how it moved so fast as there was no fin or tail movement.

The Scotts were familiar with elephant seals and other seals and sea-lions, and were sure that the creature was none of these.

Included in the report were several other sightings which could fairly reasonably be put down as elephant seals, and there was a series of second-hand sightings – those that the authors feel are less reliable, being not eyewitness reports or anonymous – going back to 1892. Two reports caught my eye, an anonymous one about an incident in 1946 and a story told second-hand about an event in 1961.

The first letter started simply: 'Back in our log, I came across a passage that might interest you.'

The writer continued by describing how the ship was crossing Marble Bay on Texada Island to Jervis Inlet. They were approaching kelp beds at Scotch Fir Point when the crew noticed something in the water. They stopped the engines.

> This thing had a triangular head which it kept moving around as it watched us. A long thin neck – we guessed 3 – 4ft, which tapered. We saw 2 coils, and guessed it to be 25 – 30ft long. It idled around watching us – when we speeded up the engine to get a little closer, it just gave a wriggle and disappeared below the

surface. We waited and it reappeared, still idling and moving its head around.

Within an hour, two people from Stillwater, and a tug-boat captain, reported the same thing.

The second report describes a creature that came out on to the land. The eyewitness was Rudi Witschi, but the authors have failed to contact him. The story, retold by Dr John Green of Memorial University, Newfoundland, referred to a time when Witschi was working with a bulldozer on the shores of the Juan de Fuca Strait near Port Renfrew. He saw a creature heading rapidly for the shore and then saw it haul out on some rocks. The two looked at each other and Witschi thought it best to retreat.

The animal he saw was large. The estimate was about 3.4m (11ft) in length, including a 1m (3 – 4ft) long tail that was flattened like a beaver's but only 10 – 12cm (4 – 5in) in diameter. The head was also large, bulbous, and less than 1m (2 – 3ft) in diameter. It had a large mouth with visible teeth, no ears, but large saucer-shaped eyes.

It had legs, not fins or flippers, that were short and thick like those of an elephant. There was no hair and the skin looked smooth. An intriguing sighting.

Since the LeBlonde and Sibert report was published there has been a steady stream of sightings and, when preparing *The Great Sea Monster Mystery,* I made contact with a couple of the more recent eyewitnesses from the Vancouver area. The first was mechanical engineer Jim Thompson who, in the winter of 1984, was mooching for Chinook salmon from his 14½-foot Folbot kayak in Vancouver's outer harbour. He was about 183m (200yds) from the most westerly Spanish Banks channel marker. It was early in the morning, it was calm, and there were no other boats about. Mr Thompson was concerned about ocean swells that might upset his canoe and, while looking about, saw a strange creature swimming in the sea about 45 – 60m (150 – 200ft) away. He went on:

It had obviously seen me before I spotted it, and it was moving at, I would say, some 12 to 15 mph out towards the Pacific Ocean.

Its head and neck were projecting above the water surface from 1½ to 3½ feet at least, and it had the upper body and head of an animal that looked like a giraffe. It had two floppy ears, one was over its face and the other one was upright. It was tan coloured with a white stripe down its brisket. It definitely had fur. It had a black snout. I'm sure I saw two stub horns.

Its head and neck craned around and looked at me specifically;

it was almost looking backwards, away from the direction in which it was travelling.

I could tell that it was making an undulating motion, in-and-out of the water, up-and-down type, like a serpent, which moved it through the water at a high rate of speed.

It had a large body, I would guess it would be in the order of 18 to 20 feet long and 1½ feet wide at least.

It disappeared in one of the swells that I was looking out for. It was looking back at me and just submerged into the swell and I never saw it again.

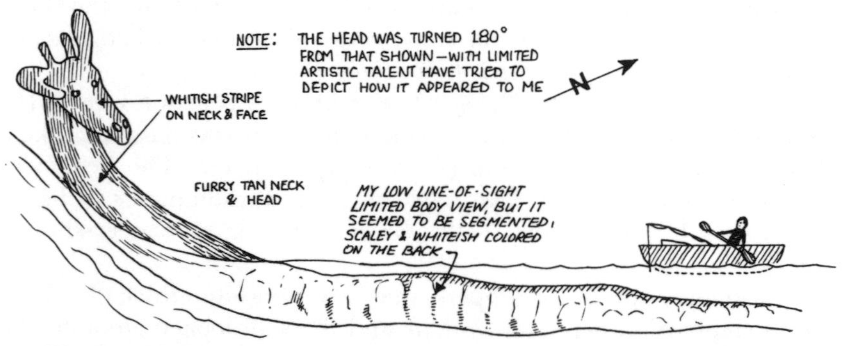

The Spanish Banks sea serpent of 1984. *From a drawing by Jim Thompson.*

The second recent eyewitness was Mr Cole who lives on the Sunshine Coast. On New Year's Day 1985, he walked to his living room window and was just about to tuck into a plate of sandwiches when he spotted something in the sea.

I looked out and against a sort of wall of mist I saw what I thought was a man standing in the bow of a boat – we have lots of 12 and 14-foot boats out here – and I thought, 'Oh that's kind of idiotic, this time of the year you can die from hypothermia,' so I picked up some binoculars – they're 7 x 50, and lo and behold it wasn't a boat, it was a creature.

I turned around to yell at the other two and put down my plate of sandwiches, and by the time I got all that sorted out, I looked back and there wasn't anything. Then I began wondering whether I had seen anything at all.

However, the amount that I saw through the binoculars, holding them with one hand, made me realize that that was a creature of some kind.

It had what I call a kind of dinosaur head; it wasn't a horse's head. It was more like those long-necked dinosaur heads, with a

> fairly small head in proportion to the body. It had a columnar neck.
>
> The body curved up like a swan's body; there was a hump that petered down to a tail. In retrospect, if I had thought it was a man standing in a boat, then the neck must have been about 5 to 6 feet tall, and the body 14 feet long with a rise of about 3 feet.
>
> It was a large chunky body, it wasn't serpentine.
>
> I am convinced that it was something that is not normally understood. I know the mammals that are around this coast. I was born in this area 73 years ago and I've been near the water all my life. I know it wasn't a seal, elephant seal or sea-lion. I thought 'whale' first, because we have the orcas with the large dorsal fin that go by in pods. I am sure, though, that it was not a normal animal.

When I came to talk to Paul LeBlonde himself, I was interested to hear of his favourite sightings. One particularly stuck in his mind:

> In the 1930s a couple of men were shooting duck on one of the islands near Vancouver. One man had shot a duck which had fallen into the water and he went to a boat in order to pick it up. Just as he arrived at the bed of seaweed in which the duck had fallen, he saw something under-water, and then this creature came out of the water – it was horse's-headish and had a neck the length of a man's height – and it swallowed the duck right in front of him. He was almost near enough to touch it. In a case like that, there is very little room for optical illusions.

Armed with the anecdotal evidence, LeBlonde and Sibert attempted to group the different animal types together and work out a simple classification, LeBlonde recalled:

> What we tried to do with the observations that we had, which were not very numerous, was to arrange them in categories that might reveal something about the animals themselves. And we had to proceed without prejudices and preconceived ideas. So, we just picked features that had been remarked by everybody, such as the length of the neck or the size of the eye.
>
> These are really artificial categories based strictly on the data available and I'm not sure that they have any zoological meaning, but I think one has to treat these rather fragmentary observations as objectively as possible.
>
> There seem to be two kinds of creature which are seen in the waters of the Strait of Georgia, and which are seen occasionally by people on the seashore or on the sea. One of them is very often

described as having a head which looks like a horse's head; some people think it looks like a giraffe, others that it looks like a camel, but you can see that all these animals have a certain morphological similarity in the form of their head – they have a long snout and ears and noticeable eyes. This animal has a rather shortish neck – somewhat less than a metre of neck.

We perceived that there might be two varieties of this animal, some with bigger eyes than others, but we're not too sure about that, of course.

The other kind is a larger creature with a neck in the two or three metres dimension, with a head again something like a giraffe or horse-like, sometimes with little horns on it.

The more common type is the short-necked one.

LeBlonde and Sibert's classification is loosely based on the observed sizes and shapes of heads, necks, and bodies; but what about behaviour? LeBlonde continued:

To my knowledge there has never been an example of an aggressive behaviour towards observers. The behaviour is usually one of an animal that has a certain degree of curiosity, but not too much. In some cases it appears to be interested in observers, and takes a quick glance at them, and then dives away again.

Asked whether he thought sea monsters existed at all, LeBlonde conceded that such a simple question could not possibly have a simple answer. He did, though, offer some personal feelings.

When I started looking at this phenomenon about twenty years ago, I was driven by a sceptical curiosity. After looking at the evidence over all these years, I am closer to thinking that they do exist than before; the evidence has revealed itself to be quite consistent and the more one sees of it, the surer one becomes that indeed there seems to be an animal that does exist.

I'm not fully convinced; you know what scientists are like – you have to have an absolute demonstration to remove the final doubt. In doubt there lies discovery.

And that demonstration will have to be a good photograph, a convincing carcass, or preferably a live animal.

To the south of the Vancouver area, the good people of Oregon are also convinced that they have been seeing live sea monsters. There is 'Colossal Claude' and 'Marvin the Monster', both creatures figuring in a newspaper report by Peter Cairns, which is

Neg No 292265 American Museum of Natural History

Ships have been known to come to an abrupt stop when they hit a harmless, filter-feeding whale shark (*above*), the largest living fish in the sea. The reconstructed jaws of megalodon (*left*), however, indicate that a far larger predatory fish once roamed the world's oceans.

Neg No 32603 American Museum of Natural History

Planet Earth Pictures – Kurt Amsler

The manta ray (*above*) is unusual amongst rays in that it sieves plankton from the surface waters. The red pogonophoran worms (*below*), together with their symbiotic bacteria, are found in the deep sea, living on sulphur discharged from hydrothermal vents on the ocean floor.

Planet Earth Pictures – Robert Hessler

attached to the LeBlonde-Sibert report. The article is dated 24 September 1967 and it appeared in *The Sunday Oregonian.*

The first encounter with Claude, according to Cairns' report, was in 1934 when the crew of the Columbia River lightship and its supply tender *Rose* spotted the 12.2m (40ft) long creature heading in their direction. According to newspaper reports at the time, it had a neck about 2.4m (8ft) long, a large bulky round body, a large tail and a snake-like head. The lightship crew wanted to lower a boat and investigate further but their captain thought it unwise.

In 1937, Claude reappeared before the fishing boat *Viv.* The skipper described a 'long, hairy, tan-coloured creature, with the head of an overgrown horse, about 12.2m (40ft) long, and with a 1.2m (4ft) waist measurement'.

One enterprising soul, fired by the size of Claude, thought to erect a sea serpent canning factory at Ilwaco. In the meantime, the creature raided the fishing lines of Captain Chris Anderson of the *Argo,* and swallowed a 9kg (20lb) halibut.

In more recent times. 'Claude' became 'Marvin' and he was supposed to have been video-taped by Shell Oil Company divers in 1963. The tape is purported to show a 4.6m (15ft) long animal with barnacle ridges spiralling along its body. It moves by a corkscrew motion. Identities, suggested by local marine biologists, range from ctenophores to salpids to siphonophores (see page 170), but Marvin's real identity is unknown.

I have tried to trace the tape, but so far without success. Marvin must remain a mystery.

The Stinson Beach Monster and Others

The most extraordinary sightings of giant sea serpents have taken place close to the US west coast city of San Francisco.

Here, like the coastal waters of British Columbia, the sea is rich with marine life. The southward-flowing Californian Current passes through an area of upwelling off the Californian coast, so nutrients from the bottom of the sea are brought to the surface layers providing food for a multitude of fish and other marine life. All along the coast there are important breeding sites for harbour seals, and Steller's and Californian sea-lions. Elephant seals have made a tremendous comeback after the bloody seal slaughters of the early twentieth century. Great white sharks, the largest predatory fish in the sea, wait offshore for the weak and the injured.

Sea otters play in the kelp beds, blue whales feed a few kilometres offshore, grey whales pass through twice each year on their way to and from their feeding and breeding grounds, and pods of killer whales track the herds on the lookout for a vulnerable youngster or two. And then there are the unknown monsters.

The first hint that there was anything of cryptozoological significance in the area was in 1976. Alongside headlines in the *Great Western Pacific Coastal Report* that announced a threat to new town developments and watershed land for sale, Bolinas minister Tom D'Onofrio declared:

> On September 30, 1976, at 12 noon I experienced the most overwhelming event in my life. I was working on a carved dragon to use as a base for a table and couldn't complete the head. I felt compelled to go down to Agate Beach where I met a friend, Dick Borgstrom.
>
> Suddenly, 150 feet from shore, gambolling in an incoming wave, was this huge dragon, possibly 60 feet long and 15 feet wide.
>
> The serpent seemed to be playing in the waves, threshing its tail. We were so overpowered by the sight we were rooted to the spot for about 10 minutes. I literally felt as if I were in the presence of God. My life has changed since.

The Agate Beach monster did not 'step out of its closet' until seven years later when newspaper publicity drew attention to a similar leviathan observed on 1 November 1983 at 2.30 in the afternoon by a road construction crew from the Californian Department of Transportation. They were on a cliff-top road – Highway 1 – to the south of Stinson Beach in Marin County, north of San Francisco. There was traffic control with two-way radio contact between the two ends of the construction area, which was about 45.7m (150ft) above the sea. The elected spokesperson for the crew was safety engineer Marlene Martin. When I spoke to her, she relived the entire experience.

> The flagman at the north end of the job-site hollered, 'What's in the water?'
>
> We all looked out to sea but could see nothing so the flagman, Matt Ratto, got his binoculars. I finally saw the wake and I said, 'Oh my God, it's coming right at us, real fast.'
>
> There was a large wake on the surface and the creature was submerged about a foot under the water. At the base of the cliff it lay motionless for about five seconds and we could look directly down and see it stretched out. I decided it must have been about

100 feet long, and like a big black hose about five feet in diameter. I didn't see the end of the tail.

It then made a U-turn and raced back, like a torpedo, out to sea. All of a sudden, it thrust its head out of the water, its mouth went towards the sky, and it thrashed about.

Then it stopped, coiled itself up into three humps of the body and started again to whip about like an uncontrolled hosepipe. It did not swim sideways like a snake, but up and down.

I had the binoculars and kept focused on the head. It had the appearance of a snake-like dinosaur, making coils and throwing its head about, splashing and opening its mouth. The teeth were peg-like and even – there were no fangs. The head resembled the way people drew dragons except it wasn't so long. It looked gigantic and ferocious.

I did not see any fins or flippers and it has bothered me that it could move so fast in that way. It was scientifically impossible for anything to go that fast without them. It was not like a snake going sideways, it went up and down.

It stunned me, never in my wildest dreams could I ever have imagined a thing to be so huge and go so fast. I thought, when I saw it, this is a myth.

There were six of us at this time all looking over the rail in disbelief. I was so glad that everybody saw the same thing!

The construction crew were rooted to the spot. A truck driver, Steve Bjora, estimated that the creature had moved at a speed of 80kph (50mph). When asked if the animal had any eyes, Marlene Martin hesitated, and then answered.

I've never really told anybody this before and I cannot swear to it but the eye that I saw looked like it was red, a deep burgundy-ruby colour. I've always hated to say that I saw that red eye. When I think about the thing I still see that red colour and what's amazing about it is that I've never seen that particular red on anything before.

After the creature had disappeared out to sea, they sat around and assessed the episode. One of the crew refused to say that he saw anything. He was concerned that his credibility would be ruined. The rest of the crew similarly decided to say nothing about their experience. But when some of the crew were talking about the day's events at a local pizza parlour, the news got out. The result was that others began to call and relate similar tales. Apart from D'Onofrio's experience, Ruth Aryon from Fairfax was visiting her daughter in Bolinas in August 1983 and saw the serpent. Marlene Martin also received a call from a Portland, Oregon, radio station about a similar creature seen off the Oregon coast and from a Pacific station where fishermen had seen it the day after the construction crew.

D'Onofrio had, by this time, finished his dragon sculpture and Martin and some others went to see it. It was, she said, similar to the beast they had seen at Stinson Beach.

Both D'Onofrio and Martin felt that the creature should be protected as it was likely that some nutcase would try to kill it. Marlene Martin told the local *Coastal Post*: 'He should have the whole damn ocean, it's his territory. He's the King of the Sea!' What the animal was doing in the shallows was not clear, although Marlene Martin had seen several sea-lions and many seabirds in the area before it arrived. After the event, all the wildlife disappeared for the rest of the day.

Seals or sea-lions were also in evidence before an even more remarkable sighting told to me by the Clark twins – Robert and William Clark of San Francisco.

It was early in the morning on 5 February 1985. They were sitting in their car just 17m (57ft) from the sea wall on the waterfront at Marina Green beside San Francisco Bay. The tide was high, the sky blue, and the visibility good. William Clark was in the driver's seat.

> I was looking directly in front when I noticed four or five seals, swimming at a fairly rapid speed, about 150 yards away. They suddenly made an abrupt turn and headed for the sea wall. Two seals were moving extremely fast. After a few seconds I saw a wake slightly to the rear of these two seals and looking closer I could see a large black snake-like animal swimming rapidly after the seals.
>
> The shape was definitely that of a snake or an eel and the head was visible as it swam perhaps only a foot under the water. Behind the head I could see what appeared to be four humps and the animal appeared to be propelling itself by wriggling in a vertical manner. A series of four coils was created at the front half of its body and these travelled backwards along the length of the neck where they would meet the middle body. At this point the undulations stopped abruptly and slowly dissipated along the remaining part of the body which was dragged behind. There would be a short pause and another series of coils would begin. What made it even more astonishing was that this was all happening at a very high speed.

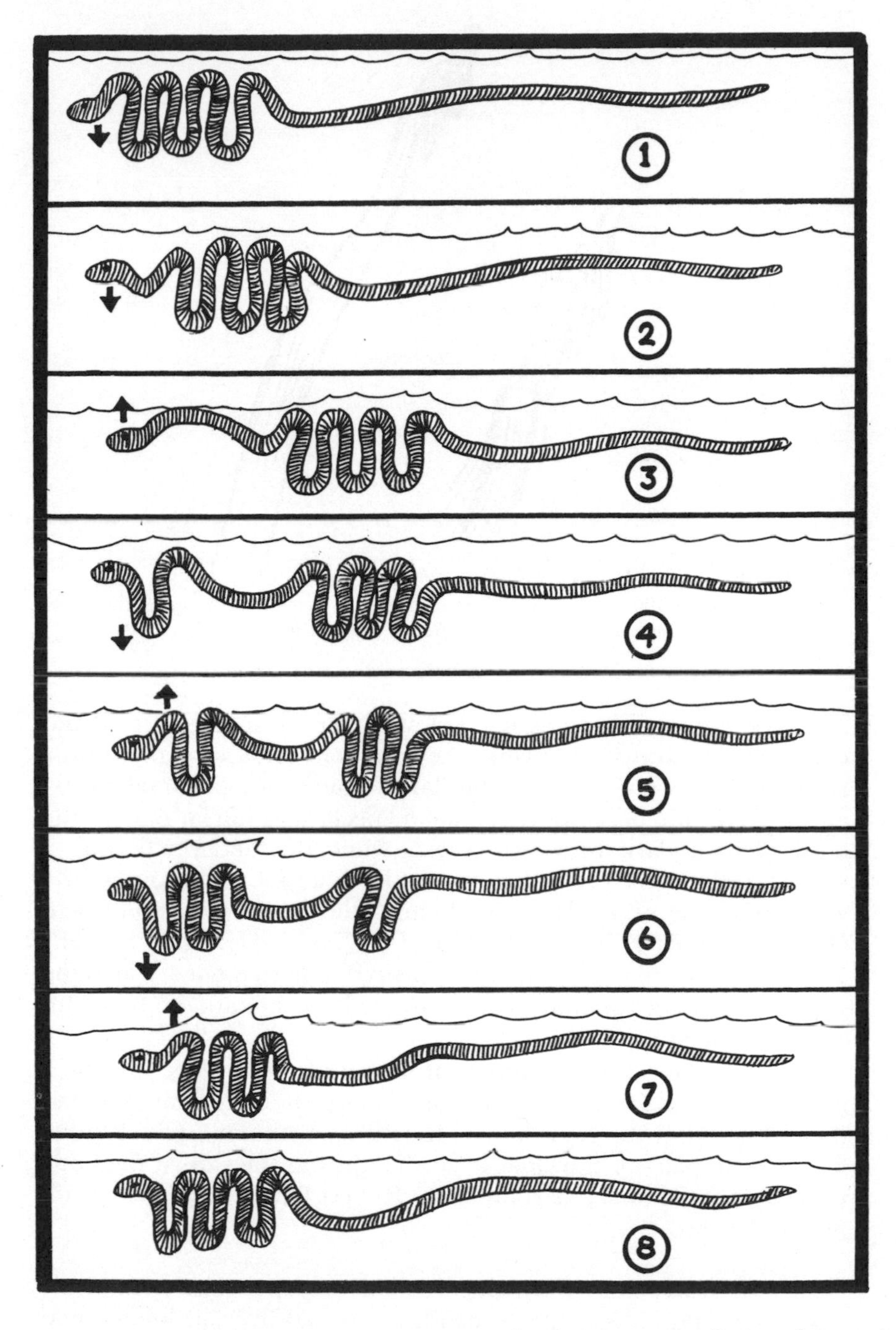

Side view of the San Francisco Bay sea serpent showing the vertical flexing of the neck portion of the body. *From sketches by Robert and William Clark.*

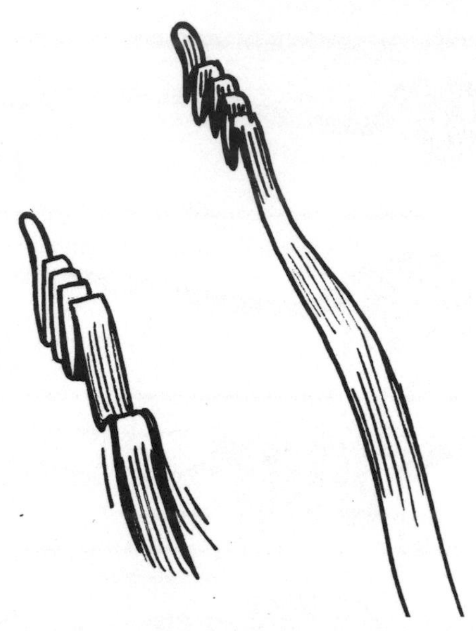

Back view of the San Francisco Bay sea serpent to show the folding of the neck region. *From sketches by Robert and William Clark.*

About twenty yards from the sea wall the animal emerged and corkscrewed around allowing the brothers to see many of its features. It was not furry but had large 'round or hexagonal scales' and looked 'oily or slimy'. The head, neck, and dorsal part of the body were a dark brownish-green, while the underbelly was a lightish yellow-green. The head was like that of a snake and a little wider than the neck. The neck behind the head was about 25cm (10in) in diameter. The main part of the body was about 1m (3ft) across and the tail was long, and appeared to flatten out towards the end. The entire animal was estimated to be about 18 – 23m (60 – 75ft) long.

It paused for a few seconds as if looking for the seals, and then twisted again so that its body was once more out of the water. It was then that the two meticulous observers, both ex-Eagle Scouts, noticed two pairs of translucent fan-like fins at the side of the body which appeared to act as stabilizers. Robert Clarke kept his eye on one of the fins.

> It was triangular in shape. It had a paper-thin membrane between the ribs. The appendage looked like it folded and unfolded like a fan, and was flexible. The membrane appeared to be a light grassy green and the ribs a mossy green.

Views of the San Francisco Bay sea serpent from the car's front window a few yards from the sea wall. It shows the rayed fins on the creature's body.
From a sketch by Robert and William Clark.

Eventually the creature, moving with undulations of the neck, headed for deeper water and disappeared. Robert and William continued to watch for a half an hour and then telephoned the Coast Guard. Later, before a Notary Public, they wrote down their observations and sent them to the International Society of Cryptozoology. Marine biologist and ISC board member Forrest Wood was understandably puzzled. In his report he wrote:

> Taking the descriptions at face value, we can identify it as vertebrate, but beyond that its features become contradictory. No known fish or aquatic reptile swims with vertical flexions of the body. Cetaceans, sirenians and true seals do flex their bodies in the vertical plane, but their flexions are modest, and they are quite incapable of forming humps, or 'coils', such as those described.
>
> The scales, slimy appearance, and pairs of rayed fins could belong only to a bony, ray-finned fish (except that I'm not aware of any fish which the fins are attached to the body along one edge). With the exception of sea-horses and some of their relatives, no fish has what could be called a neck. In all cases, cervical vertebrae are lacking; neck movement is impossible.
>
> In any case, the described creature cannot be assigned to any class of vertebrate. On the basis of zoology, including paleontology, and phylogenic principles, it is an impossibility.

For an identity, I guess we just have to wait for the next appearance of the 'impossible' animal.

4

Serpents of The Seven Seas

'Seafaring men expect storms and sometimes wrecks,' wrote Joseph Ostens Grey, Second Officer of the SS *Tresco,* 'but for most men of the merchant marine in times of peace there is much monotony in their voyages to and from the various ports they seek during their years at sea. On an ordinary voyage, such as I have taken, year in and year out, for sixteen years, a remarkable experience befell me recently.'

The article appeared in *The Wide World Magazine,* and Second Officer Grey was referring to an incident that took place on the morning of 30 May 1903, about 145km (90mi) south of Cape Hatteras. The sea was calm and the sky bright. Grey was on duty on the bridge.

At about 10 o'clock in the morning, the ship encountered considerable commotion in the water caused by a school of what Grey described as 'the usual bottle-nosed sharks' that were swimming 'shoulder to shoulder closely packed together'. Grey counted about twenty sharks in the group. Then, about an hour later, he spotted something else in the water, due south-east of the ship – the same direction from which the sharks had come.

Both Grey and the quartermaster at the wheel thought it to be an abandoned hulk. Grey ordered the ship to change course to come closer to the derelict – it was the duty of all ships to report such obstacles in the sea to the Hydrographic Office – but as they got closer they realized that not only was their ship moving towards the object, the object was also moving towards the ship. Grey wrote:

> With a conviction that grew even deeper, and ever more disquieting, we came to know that this thing could be no derelict, no object the hand of man had fashioned, no object, the eyes of man had ever seen.
>
> Now, swiftly, with terrible uprising, a mighty and horrible head came out of the water, surmounting a tall, powerful neck that had the thickness and strength of a cathedral pillar, yet was spindly in comparison with the huge and awful head that it supported . . . I felt cold, with an unknown, overmastering horror. But I stood at my post on the bridge looking at that

> immense, dragon-like head, reared high on the long, powerful neck. It must be a sea serpent – what else could it be? – coming towards us, and now so near.

Pandemonium broke loose on the ship. Everybody cleared the decks and shouted to Grey and the quartermaster to get below and steer the ship from the wheelhouse. This they did, and with great haste, but Grey continued to watch the creature from one of the chartroom portholes. They were terrified that the creature would attack the steamer. The ship was running without cargo and was trimmed with ballast. Any substantial weight placed on either side of the deck would have caused the ship to list and perhaps to have turned over. The ship's rail was 6.1m (20ft) above the sea surface. The monster's head was at 4.6m (15ft) and could have risen further. There was a danger it might try to clamber aboard. Grey continued:

> There that evil thing remained, the body motionless, the tail undulating vertically. As it lashed the water with the long, snake-like tail the head all the time was reared high, regarding the *Tresco* as if waiting to see what such a thing as a ship might be and, until it should decide, determined to maintain its watchful position. It looked for all the world like some fantastic Chinese dragon become a living reality; or a page from a scientific work picturing some ancient saurian monster, neither reptile nor beast wholly, but both in part.

He was quick to point out that seafarers are trained to observe, and this creature was so close – just two ship-lengths away – that they could note its size, shape and features in some considerable detail.

The body was estimated to be about one-third the length of the ship – about 30.5m (100ft). The widest part of the body was about 2.4m (8ft) across, giving a circumference of about 6.1m (20ft). The body was not completely cylindrical but was arched towards the top, and the back humped down to the neck and tail. The widest point was at the forward part of the body, tapering towards the tail.

The head was about 1.5m (5ft) long and the neck about 46cm (18in) in diameter. The snout, devoid of nostrils or blowhole, was upturned but blunt at the tip. The lower jaw protruded further than the upper. Under the lower jaw there seemed to be a pouch or flap of skin.

> Presently I noticed something dripping from the ugly lower jaw. Watching, I saw that it was saliva, of a dirty drab colour, which dropped from the corners of the mouth.

> While it displayed no teeth, it did possess very long and formidable molars. There were two, curving down and backward like walrus's tusks, about eighteen inches in length, at the very back of the mouth. They were of a dirty ivory hue. If it had teeth or tongue it did not show them; but we saw that its mouth was red.
>
> Its eyes were also of a decided reddish colour. They were set high in the head, like serpent's eyes, or those of water-fowl. They were elongated vertically, not lateral, were slightly elliptical rather than round, and were large in proportion to the head. Their greatest length was about seven inches and their extreme width four inches. No pupil was visible. The entire cast of the eyes appeared to be red, of the shade of maroon. They carried in their dull depths a sombre, baleful glow, as if within them was concentrated all the fierce menacing spirit that raged in the huge bulk behind.

Grey went on to describe small 8cm (3in) scales below the eyes which 'dragged backward'. They became increasingly large behind the head until they were the size of 'great plates' on the body itself. He considered the colour of the beast to be 'antique bronze' with the light green colour of oxidized metal. Grey continued with the visible appendages.

> Its side fins, extending one-third of the way from the shoulder to the beginning of the tail, and broadest – about a foot – near the shoulder, worked like fans in swift agitation of the water.
>
> As I gazed, fascinated with the horror of the thing, it raised its dorsal fin, obviously in wrath. And then a thing happened which, strange as it may appear after the recounting of the fearsomeness of the serpent's dreadful front, was more appalling, more sickeningly terrifying, than anything I had yet beheld. Suddenly, at the back of the head, a great webbed crest uprose, and from the eyes, hitherto so dull save for the glow smouldering in their depths, a scintillating glare appeared, as if the creature felt the moment had come for attack. The crest was a foot in height at its forward extremity, where it was supported by a sharp-pointed spine.

The creature then, according to Grey's account, churned up the water with its tail and was clearly ready to pounce, but it hesitated, not knowing what to make of the ship. To the crew's relief, it began to turn away.

> Its great body turned, as if on a pivot, inward in a circle, followed

> by its long tail. With astonishing ease for so huge a bulk it made the sweeping evolution. And only then did it lower its ugly head, that had so long confronted us in open antagonism.

At this point Grey realized he had not informed the captain; he raced to his stateroom, burst in, and blurting out, 'Come on,' Captain, quick, come and see this animal.'

The captain sprang up from his bed where he had been resting, and joined the mate on the poopdeck. The monster was, by this time, about 400m (¼ mi) away. The captain, Captain W H Bartlett from Looe in Cornwall, stood transfixed.

> 'Good heavens! What's that?'
> 'I take it, sir,' I replied, 'to be a sea serpent.'
> 'I believe you're right,' he rejoined.

Eventually, the creature disappeared from view and the ship slowly got back to normal. When it reached port in Santiago, in Cuba, Grey prepared a report which was signed by all those of the crew who saw the creature (some were asleep at the time).

It was, without question, a remarkable story enhanced, no doubt, by an editor with a vivid imagination. Heuvelmans is dubious about the whole story. He considers it to be a complete invention, although he does acknowledge that the *Tresco* is a real boat and the details about it, checked with *Lloyd's Register*, are valid, 'The writer,' declares Heuvelmans, 'clearly took more trouble with his shipping information than his zoology.'

Reference to the *Tresco* sea serpent, however, does appear in the ship's log; part of the entry for Saturday 30 May 1903 reads:

> 10am Passed school of sharks followed by a huge sea monster.

It was signed by the Chief Officer, Mr Griffiths, who was one of the few people on board who had been fast asleep and had missed the entire episode.

Was it, then, a fabrication – a means of gaining attention and money, or did the crew of the *Tresco* have the rare privilege of seeing one of Nature's mysterious creatures? We shall never know.

But the crew of the *Tresco* have not been alone. Down the centuries, sea serpents have been observed and logged in most of the seas in the world. The most famous encounter, perhaps, was that of the *Daedalus*.

The Royal Navy frigate HMS *Daedalus* was returning to Plymouth after a tour in the East Indies. The year was 1848, and on 6 August the ship was at latitude 24° 44′ south, longitude 9° 22′ east, heading north-east on a line between the Cape of Good Hope and the island of St Helena.

At five o'clock in the evening, a midshipman drew the Officer of the Watch's attention to an object in the sea. He was on the quarterdeck with the Master and the Captain. The boatswain's mate and a helmsman were at the wheel. The rest of the crew were at supper.

The *Daedalus* sea serpent of 1848. From an engraving in *The Illustrated London News.*

As they got closer, they could make out that the object was a sea serpent with its head about 1.2m (4ft) out of the water. They estimated it to be 18.3m (60ft) long, with no visible means of propulsion. It did not undulate – horizontally or vertically, but still managed a fair rate of knots, in fact, about 19 – 24kph (12 – 15mph) in a south-westerly direction. It came so close to the ship that those assembled on the deck could see some detail.

The neck portion was about 38cm (15in) behind the head, which resembled that of a snake. The colour was dark brown, with a yellowish-white on the throat. There was a seaweed-like mane on its back.

When the ship returned to its home port, news of the encounter got out and the Captain was asked by the Admiralty to confirm or deny the rumours. This he did, together with a letter to *The Times*. Some of the ship's officers offered stories to other newspapers and magazines.

'It was just an enormous strand of seaweed,' declared two ship's captains who had seen giant kelp adrift in the South Atlantic in the same spot as the *Daedalus* encounter.

'It was most definitely animate,' replied a *Daedalus* officer. The debate continued for many years afterwards and the *Daedalus* affair is still the most quoted of all sea serpent sightings.

Another naval officer in another war at another time was forced to make a similar statement. The year was 1915 and the place, the North Atlantic. On 30 July, the German submarine *U 28* torpedoed the British ship *Iberian* to the south-west of Ireland. It sank rapidly, stern first, and went towards the bottom of the deep sea. After about twenty-five seconds, the crew of the submarine heard a loud explosion, and in a fountain of seawater and wreckage, that was shot to a height of 30m (100ft) in the air, there was a 'gigantic sea-animal'. The creature, which looked like a 18.3m (60ft) long crocodile with four limbs and webbed feet, writhed and struggled at the surface and, after fifteen seconds, sank below.

Three years later, another German U-boat, *U 109,* was in the North Sea when, at ten o'clock in the evening of the 28 July 1918, the captain and a member of the crew saw a 30m (100ft) long creature with jaws like those of a crocodile. It also had legs with 'very definite feet'.

The *Fly* sea serpent. *From a drawing by Carton Moore Park.*

Crocodile-shaped sea monsters, it seems, are not uncommon. The captain of the steamer *Grangense,* on passage between New York and Belém near the mouth of the Amazon River, saw a creature with a crocodile-like head apparently playing at the surface. Its jaws contained rows of regular 10 – 15cm (4 – 6in) long teeth.

And in the late 1830s, the British ship HMS *Fly* was on duty in the

Gulf of California when the captain spotted, through a calm and transparent sea, a 'large marine animal, with the head and general figure of the alligator', except the limbs were two pairs of 'flippers', like those of sea turtles, rather than legs. It was chasing prey.

In more recent times, yachtsmen and women have been going to sea in even smaller ships and boats than the mariners of old. Those crossing the Atlantic in rowboats or sailing it single-handed are inevitably closer to the sea surface than ever before and in a position to observe the wildlife that drifts or swims by. British explorer and yachtsman John Ridgeway is one of those who has seen something unusual in mid-ocean. In the summer of 1966, he and yachtsman Chay Blyth were rowing across the Atlantic, from Cape Cod to Ireland, in a small 6.1m (20ft) long rowboat. They had just 46cm (18in) of freeboard, so were about as close to the ocean surface as one can get. One dark July night, when Blyth was asleep in the stern and Ridgeway was rowing alone, a strange and scary thing happened. Twenty years after the encounter, during the preparations for *The Great Sea Monster Mystery,* he recalled, in a down-to-earth way, what happened.

> There was no moon, as I remember it, and only a gentle swell. I was facing the stern of the boat, rowing in the bow position with two oars, and looking out to my left I saw a luminous track racing directly towards the boat.
>
> I was startled by this. It looked not unlike a torpedo coming towards the boat. For an instant, I thought something was going to hit us, but it disappeared under the boat, came up on the other side with a hissing sound, and then it disappeared into the dark.

In their book *A Fighting Chance,* Ridgeway's description is a little more colourful:

> I was shocked into full wakefulness by a swishing noise to starboard. I looked out into the water and suddenly saw the writhing, twisting shape of a great creature. It was outlined by the phosphorescence in the sea as if a string of neon lights were hanging from it.
>
> It was an enormous size, some thirty-five or more feet long . . . It headed straight at me and disappeared right beneath me . . . I forced myself to turn my head to look over the port side. I saw nothing, but after a brief pause I heard a tremendous splash.
>
> I thought that this might be the head of the monster crashing into the sea after coming up for a brief look at us.

In the book, Ridgeway states that he searched for a rational explanation. Both he and Blyth had seen whales, sharks, dolphins, porpoises and flying fish, but he felt that the creature they had met that night was none of these.

> I reluctantly had to believe that there was only one thing it could have been – a sea serpent.

In conversation in 1986, with twenty years' hindsight and having clocked-up many long voyages around the world, John Ridgeway reconsidered his interpretation.

> I came to the conclusion that this was a porpoise or dolphin swimming fast, close to the surface, disturbing the luminous trace in the sea and making this fire-like, 'snaky' approach to the boat; the 'snaky' effect created by the swell. It was a very dark night, and I could imagine that, say, in the Middle Ages, when people would have been very religious or very superstitious or both, thinking they were about to sail off the edge of the earth, and were told that there were sea monsters, they would easily believe that this luminous wavy track was a sea serpent.

On another occasion, Ridgeway was sailing around the bottom of the world in a 15.2m (50ft) long yacht and had travelled for about five months with few visits to land when he was in the Pacific and approaching Cape Horn. After a long period in blue waters and a spell in dense fog, ahead of the ship were huge seas and black ragged-topped cloud. At this point, with the crew in a highly charged and emotional state, some creature came up behind the yacht. The crew had seen albatrosses, whales, and luminous squid at night but again this was not like any of those. Ridgeway, himself, lives in Scotland and sees seals virtually every day he is there, but this creature, at first glance, did not look like a seal.

> The boat was moving at about 9 or 10 knots, which is pretty fast for an animal to keep up with for a long period.
>
> It looked like a seal but wasn't a seal. I came to the conclusion it was a sea-lion. It was not the bluey-green of the sea, it was a yellowy-brown colour, and moving with considerable 'sinuousness'. It was very athletic, a long way out at sea, and moving at great speed through quite large waves, and appearing now and then. It was coming through the sea with a sharp, pointed head which, I suppose, if it was continued, could seem to be the front part of a sea serpent.

One might think that people in small sailing boats would be seeing unusual or unidentified sea creatures with unfailing regularity. Sailing boats do not disturb the marine environment unduly and do not scare timid animals away. Indeed, sailing boats have been used recently as observation platforms from which to study the behaviour of sperm whales in the Indian and Pacific Ocean. It is surprising, then, that there are so few sea serpent sightings from small yachts. There was, however, one interesting case involving a creature with the unlikely name of 'sea monkey'.

Distinguished British yachtsman Brigadier Miles Smeeton was sailing the Aleutian Islands in the North Pacific in 1965. He was aboard his 46-foot ketch *Tzu Hang* together with his wife, daughter, and a friend, Henry Combe. They were about 6.4km (4mi) off the north coast of Atka when a strange animal appeared. Smeeton estimated it to be about the size of a sheep with long reddish-yellow-coloured 'pepper-and-salt' hair, like that of a Cairn terrier but about 13 – 15cm (5 – 6in) long. The hair floated about the body much like seaweed on a rock. As the boat approached, it dived. Smeeton's daughter saw it surface and described the head as more like that of a dog, with eyes set close together.

Henry Combe confirmed the description in conversation with Miles Clark who wrote an article about the 'sea monkey' in *BBC Wildlife Magazine.*

> It was misty, but clear around *Tzu Hang* that morning. The sea was flat calm. I saw the animal from a distance of, at most, 10 feet. It was lying in the water with its head cocked up looking at me. My mother had a Tchitzu Tibetan terrier at the time, and I remember I was struck by the resemblance. The head seemed very small in relation to the body but that may have been because all the hair was floating out from it.
>
> The hair was, as I remember, sort of palish grey with a mangy tinge to it. It formed a distinctly dark, straight 'parting' all down the animal's back. It's strange but I cannot really recall its size – my impression seems to be that it was about five feet long but I feel somehow that it was less. I think it just appeared much bigger because of all the surrounding hair.
>
> I also remember its dark intelligent eyes. I saw it for about 10 to 15 seconds. I never saw any legs, feet or fins. We saw seals enough for me to state quite categorically that what I saw was no seal. One thought which has occurred to me was that it might have been a sea otter with a strange coat, perhaps a mutation. Perhaps he had something wrong with him, perhaps he was very, very old, I don't know. I saw lots and lots of sea otters and

certainly this was very different but, perhaps . . . who knows.

In 1978, a hairy red-coloured sea creature had been seen near Unalaska by a fisherman. On closer inspection he was able to identify it as a sea otter. But the most interesting reference was in 1741, when the fortunate observer was none other than Georg Wilhem Steller, the young naturalist who sailed the northern seas with Vitus Bering and who lent his name to many marine animals – Steller's sea-cow, Steller's sea-lion, to name but two. His ship was south of the Semidi Islands on 10 August, when he spotted a strange animal in the sea. He wrote:

> It was about two Russian ells (1.5m or 5ft) in length and the head was like a dog's, with pointed erect ears. From the upper and lower lips on both sides whiskers hung down, which made it look almost like a Chinaman. The eyes were large; the body was longish, round and thick, tapering gradually towards the tail. The skin seemed thickly covered with hair of a grey colour on the back, but reddish-white on the belly; however, in the water, the whole animal appeared red like a cow.

Steller, whose description is remarkably similar to that of Henry Combe, also did not see feet or fins. He did note, however, that the creature jumped gracefully and swam about the ships for over two hours. Occasionally it would bring the front third of its body right out of the water, and when it saw some seaweed float past it grabbed it in its mouth and swam to the ship, 'monkeying' around as Steller described it – hence the name 'sea monkey'.

In seeking an explanation, I spoke to seal expert Dr Shiela Anderson, from the Sea Mammal Research Unit in Cambridge. She mentioned the possibility of a wayward Hawaiian monk seal. When this species moults, the fur around the snout sometimes forms a 'Chinaman's moustache'. It does not normally have very long hair, however, and it does not show external erect ears. It is considered by some biologists to be a 'living fossil' because of the bone structure of the inner ear; it is more like that of primitive seals that lived 14.5 million years ago than most other modern seals living today.

Fur seals and sea-lions have external ears but they are usually small scroll-shaped protruberances, of which the New Zealand fur seals have the longest ones. The eared seals did, however, evolve from dog-like ancestors about 25 million years ago. Could the 'sea monkey' be another primitive species but, unlike the monk seal which is threatened with extinction because of man's activites, could it have eluded the curious gaze of all but a handful of humans?

In the tropics, the Persian Gulf, the Indian Ocean, and the South China Sea featured in three as yet unpublished sea serpent sightings.

The first was in 1937 at about the time the Sino-Japanese War resumed. Alfred Peterson, twenty-four at the time and a male nurse, was on board a troopship which was racing across the South China Sea to relieve the British Concession in Shanghai. He was jogging around deck before breakfast but was stopped in his tracks by something he had seen on the surface of the sea.

> I looked out and I saw what I thought was a very big tree floating in the sea. A bit later, I stopped on the deck again to get a few minutes' breath and noticed that the thing was still level with the ship, and I thought to myself, 'That's not a tree, that's moving with us; a tree couldn't be moving like that.'
>
> And so I watched it. I could see the body was about 24 to 25-foot long, with a tail on the end, all on top of the water. It was a grey-black, hippo-coloured.
>
> It sort of half-turned, and its neck stuck up like a giraffe's. On the top was this giraffe head sort of shape, and I could see two sort of ears or horns or whatever; they were like drumstick ears.
>
> It simply played as it went along; it disappeared, came up again, played, and went on. It didn't look vicious; it wasn't splashing; and it looked a big, really gentle thing.

The second event was in 1938. Captain Kingston Lewis was then one of the crew aboard the tanker *British Power* on passage in the Persian Gulf between Durban and Abadan. It was a hot day and the sea was calm, and Captain Lewis was taking a siesta on the poopdeck when his attention was attracted by something in the sea.

> I happened to stand up and I was looking over the side and this animal or monster popped its head out of the water. It had a longish neck about six to eight feet long with a head on it something like a horse.
>
> It looked at the ship as if to say, 'What is this?' and after twenty or thirty seconds it ducked its head down into the water and disappeared.
>
> It had a rather large body underneath but I couldn't see exactly how big it was.
>
> I told the second officer about this and he sort of laughed it off, he didn't believe me, but not long afterwards he came down and said to me, 'You know, Lewis, you were right about that.' He said, 'It's here in the *Persian Gulf Pilot'* – the sailing directions for the Persian Gulf – and apparently a similar animal had been

> seen by one of His Majesty's survey ships, and they had sighted a sea animal or monster and had made a sketch of it.

The third sighting took place during the Second World War, in 1942 to be precise. Mr Welch was on board a troopship bound for Bombay from Durban. He was on look-out duty.

> We never knew quite what we were looking out for but we were always on the look-out.
>
> On one occasion, though, I could see a large black object way in the distance. My heart went down to my boots because I thought it was a submarine. I sounded the alarm, bells rang all over the ship, and everybody was going mad, panicking.
>
> One of the duty officers looked through his binoculars and said, 'Oh no, it's not a submarine, I don't know what it is, probably something just floating in the water.'
>
> Anyway, as the ship got nearer we could see what I can only describe as a sea monster; it was definitely something swimming. It crossed our bows and we could see it quite clearly; it was a sort of a serpent, about 20 to 30 feet long, very thick – probably as thick as a tree-trunk, and its back was arched in several places.
>
> I couldn't make out its head with any clarity; it was sort of surrounded by waves.
>
> We carried on, and it went its way while we went ours; it took no notice of us whatsoever, and eventually it disappeared from view.

A sea serpent sighting that failed to find its way to those who chronicle such things until quite recently, came to light as a result of a casual conversation between Paul LeBlonde and Professor Jean Paelinck, an economist at Erasmus University in Rotterdam, The Netherlands. LeBlonde reports the event in the Winter 1983 edition of *Cryptozoology,* the journal of the ISC. It turns out that Professor Paelinck's grandfather, Captain J Koopman, was a distinguished Dutch merchant seaman who wrote up his memoirs and included a story about a sea monster encounter in the South Atlantic. It happened in 1906, off the coast of Brazil, near Pernambuco, now known as Recife. And the story is all the more intriguing because another sea serpent sighting, made in the previous year, occurred in roughly the same spot.

The 1905 sighting was reported and analysed in the *Proceedings of the Zoological Society of London,* and was made by two Fellows of the Society, E G M Meade-Waldo and M J Nicoll on Lord Crawford's yacht *Valhalla.* They were engaged in a scientific cruise off the

Brazilian coast, and about 24km (15mi) off the mouth of the Paraiba River they spotted something unusual. Nicoll thought it was the fin of a large fish. Meade-Waldo wrote:

> I looked and immediately saw a large fin or frill sticking out of the water, dark seaweed-brown in colour, somewhat crinkled at the edge. It was apparently about 6 feet in length, and projected from 18 inches to 2 feet from the water.
>
> I got my field-glasses on to it (a powerful pair of Goerz Trieder) and almost as soon as I had them on the frill, a great head and neck rose out of the water in front of the frill; the neck did not touch the frill in the water, but came out of the water in *front* of it, at a distance of certainly not less than 18 inches, probably more. The neck appeared about the thickness of a slight man's body, and from 7 to 8 feet was out of the water; head and neck were all about the same thickness.
>
> The head had a very turtle-like appearance, as had also the eye. I could see the line of the mouth, but we were sailing pretty fast, and quickly drew away from the object, which was going very slowly. It moved its head and neck from side to side in a peculiar manner; the colour of the head and neck was dark brown above, and whitish below – almost white, I think.

Several years later, Meade-Waldo gave more information about the encounter to Rupert Gould, who wrote the celebrated book on unusual sea creatures, *The Case for the Sea Serpent,* in 1930. He mentioned that the neck made a wave as it passed through the water, and that he saw a large body under the water behind the neck. The neck was also 'lashing the sea into foam'.

The *Valhalla* sea serpent of 1905. *From an engraving.*

Nicoll wrote that he thought the creature to be 'for the want of a better name . . . the great sea serpent'. He drew a sketch of the *Valhalla* sea serpent, as it became known. The creature was debated in serious academic circles, the encounter included in books and articles, and then no more was heard until Professor LeBlonde got talking to Professor Paelinck about the latter's grandfather.

Captain Koopman's 1906 encounter was on a voyage from the Mediterranean to Montevideo. The ship was about 75km (40mi) east of Recife when Koopman saw a sailing ship with its sails flapping. He ordered the ship to alter course so that the runaway ship could be identified and reported. Then suddenly, the helmsman drew his attention to an object on the starboard side of the ship.

> I saw, about one hundred metres away, obliquely on starboard, an enormous beast whose length I approximated at about 60 metres. It was overtaking our ship, which appeared to be standing still, with the speed of an arrow off a bow. With the help of my telescope, I could form some idea of the monster, although only in an approximate fashion. The monstrous head and a number of enormous dorsal fins sticking out above water level, as well as its wide wake, showed the nearly horizontal posture of this sea-dragon or serpent.

At the same time, looking towards the Brazilian coast which was, in fact, over the horizon, Koopman could see Recife, but upside down! It was, in his words, 'an impressive *Fata Morgana'*, a mirage. At the time, he reported the mirage, which was scientifically understood, but chose to keep quiet about the sea serpent 'for fear of ridicule'.

LeBlonde worked out the co-ordinates of the two ships – the *Valhalla* was approximately 7° 14′ S, 34°25′ W, while Koopman's ship was at 8°06′ S, 34°13′ W. From the co-ordinates, it was clear that the two boats had travelled through the same piece of ocean, and it might have been that they both saw the same sea serpent.

Another encounter was uncovered during an edition of BBC Radio's long-running series *The Brains Trust* and reported in *The Listener* dated 12 June 1941. The panel, who included Julian Huxley and Commander A B Campbell, were presented with the question: 'Will the Brains Trust accept the evidence for sea monsters?' Commander Campbell was first to speak.

> I think without doubt there are such things as sea monsters. I've not only seen certain of them but I've smelt them. We were lying in a very secluded bay, north of the American coast, and taking stand-easy on deck, and suddenly a terrible smell pervaded the

> ship. I thought something had gone wrong, got up, and there about seventy yards away were three big lumps in the water, moving, and we counted along from one stanchion to another, seventy feet of this creature, and it stank most horribly. It came near to the ship, and then took a dive.
>
> The following day two of them turned up, and we could see seventy feet of them. They certainly were some sort of sea monster. Unfortunately we got no snaps of the head, but we got a photograph of the bumps. It was some huge creature without doubt.

Julian Huxley then brought proceedings back to earth with some logical explanations for the strange creatures people had seen.

After a BBC broadcast in *The Northcountryman* series yet another sea serpent came to light. In the programme, Mr J S Colman of the Port Erin Marine Station, Isle of Man, talked about the existence or not of sea serpents, with particular reference to R T Gould's book. The broadcast prompted G Cooper from Budleigh Salterton in Devon to write to *The Listener* on 5 March 1953. He recalled an encounter he had had many years earlier.

> On October 26th 1937, while approaching St. Thomas in the Virgin Islands, on board the M.S. *Amerika* of the Danish-American Line, a creature similar to those described in the discourse was seen by several passengers, including myself, who were standing alongside the rails.
>
> The weather was bright and clear and at a distance of about a quarter of a mile, travelling on the opposite course, an elongated mass, sixty to eighty feet in length, broke surface in a flurry of foam and spray. It had a long serpentine neck thrust upwards, and a flattish head which, however, was flexed and gave it an equine appearance. Behind the neck a series of six or more large, protruding humps, chocolate-brown in colour, moved forward in sinuous motion, giving an illusion of speed, which was probably not more than fifteen to twenty knots. The sea alongside was churned white, and a creamy wake followed for some considerable distance.

Mr Cooper went on to describe how he had put the details in his diary and obtained confirmation that his notes were accurate after showing them to his wife and other witnesses on the ship. His story became another piece of anecdotal evidence to be included in the long history of sea serpents. No doubt, from time to time, it will be examined and evaluated along with all the other stories. But it is clear that the only way that the scientific community will accept the existence of sea serpents and other mysterious creatures is when an intact body presents itself to be formally examined. In the meantime, I wonder how many more encounters with strange marine creatures have yet to be made public? Such a story may provide us with the one clue that we have been looking for in our attempts to solve the mystery of the Great Sea Serpent.

5
The Big Blob

Bermudan fisherman John (Sean) Ingham had many surprises while searching for new deep-water fishing grounds off the north-west coast of this group of volcanic islands in mid-Atlantic. Ingham's speciality is crab fishing, and during the early 1980s he was experimenting with crab traps set at much greater depths than usual, working them along the deep seabed and gradually up the slope of the Continental Shelf. He started at 3,475m (11,400ft or 1,900 fathoms), but it was at 1,463m (4,800ft or 800 fathoms) that things began to happen.

First he found a rich source of cod. This is Bermuda's national dish, but the species has declined so much in recent years, mainly by over-fishing, that the islanders now import most of what they eat – and so this discovery alone meant that Ingham's experiment had paid off.

Then there were traps full of huge red crabs. One crab weighed in at 7.25kg (16lb). It was as good to eat as most of the deep-sea crabs taken around North American shores, like the Alaskan king crab, but the identity remained a mystery until someone noticed the resemblance between Ingham's crabs and the golden crab *Geryon fenneri* of Florida. After a period of more intense research, two species were recognised - *G. fenneri*, the American species, and a brand-new species, *G. inghami* - by crustacean experts Raymond Manning of the Smithsonian Institution in Washington and L B Holthius of the Rijksmuseum van Natuurlijke Histoire in The Netherlands. The new species was named after the fisherman who found it and was deemed to be similar to *G. gordonae*, a crab found on the eastern side of the Atlantic and described by R W Ingle of the British Museum (Natural History), and itself only discovered and named in 1985.

Ingham's theory, based mainly on the observation that his baits were deteriorating much more rapidly that they ever had before, was that he had come across communities of deep-sea crabs associated with hydrothermic vents – hot springs that support the only life on earth that is totally independent of sunshine (see page 8).

Whatever the explanation for the abundance of crabs, he was enjoying something of a bonanza – until the day, he recalled to me, when he was hauling up a trap filled with crabs and the winch suddenly began to run backwards.

> Using water to stop the rope burning, I managed to stop it on the capstan roller. We started to haul again, and as the trap reached about 200 fathoms, we would see on the sonar that there was something very large on it. Then there was a series of hard jerks, the line broke and the trap was lost.

Ingham and his partner were shaken, but not enough to stop fishing their new rich grounds. Still, it was clear that their traps had been discovered by something, and on several occasions they felt a tugging on the other end of the line – sometimes the pull they measured at 2,700kg. Then finally came the day when they were persuaded to start fishing elsewhere, anywhere.

> I was trying to find ways of getting the trap back without damage, so I snuck the rope up tight, not trying to force it off the bottom. It was quite calm that day, with no swell, and so I was able to manoeuvre the boat right above the trap. Then we switched off the engine and waited to see what would happen.
>
> Shortly after, the boat began to move parallel to the shelf. It was towed for about a third of a mile. We had seen a large shape close to the trap and at first thought it was a large mass of crabs moving towards the bait, but when the boat started to move I knew it couldn't be the crabs.
>
> We were over a rocky bottom and could see this thing on the sonar moving from rock to rock – and then it released the trap. When we got it up, the trap was crushed a bit. I said to my partner, 'This is ridiculous. There's no sense in trying to make a living like this. Let's cut our losses and move somewhere else and see if we can avoid this thing down there.'

The crabs are thought to be preyed upon by conger eels, but whatever had designs on Ingham's crabs was of another dimension entirely. Bermudan fishermen often return with chunks of giant squid (see chapter 6), the debris of violent encounters between squids and sperm whales. But it is thought squids normally stay in mid-waters and are not bottom-dwelling animals. The cephalopods that live on the bottom of the sea and move from rock to rock are more usually octopuses. The trouble is, giant squids are known to science – at least they are today – but as far as the taxonomic establishment is concerned, there is no such thing as a truly giant octopus.

The largest known octopus is *Octopus delfei,* the Pacific giant octopus that inhabits the rocky Pacific coast of North America, particularly Oregon, Washington and British Columbia. The record size is held by an individual that was wrestled to the surface by skin-diver Donald Hagen in 1973. It had a radial span, from arm tip to arm tip, of about 7m (23ft).

Other specimens have been reported up to 11m (36ft), but these are unsubstantiated. They are considered to be relatively harmless, avoiding man and other predators by hiding in crevices among rocks. Indeed, since 1967, in the area of Tacoma not far from Seattle, there have been octopus-hunting contests in which these giants are brought to the surface, weighed and then returned to the sea. On one occasion, the contest was held near Port Townsend where many large logs were placed in the water. These large octopuses, much to the horror of the contestants, were reputed to hang on to the undersides of the logs and drop upon unwary skin-divers!

No competitor, as far as I am aware, suffered the indignity of having to be rescued from a marauding octopus. They are, it seems, tamed with relative ease by holding them at arm's length and rocking them from side to side. As long as one is gentle the arms go limp, and the creature can be gently 'wrestled' to the surface.

They are, however, hard to find. They live in crevices and crannies in rocks and are masters of camouflage, changing colour to suite their background. They are also predictable. Pacific octopuses have been known to live in the same hole for several years.

Their docility was once put to the test when a large 52kg (115lb) Pacific octopus was coerced into taking part in an edition of the television series *Sea Hunt* starring Lloyd Bridges. It became so tame that it could be positioned in the 'set' by leading it around by a tentacle. It lived happily in a rubber dinghy, consuming crabs, between 'takes'.

Do not, however, think that these animals can be taken for granted. Though divers often catch them and their relatives in other parts of the world without being bitten by the powerful, horny beak at the centre of the eight radial arms, sometimes they have been known to attack and even kill.

In October 1877, the *Weekly Oregonian* reported a North American Indian woman going to the sea to bathe and being caught by an octopus, held under, and drowned. Her body was discovered, still in the deadly embrace of the octopus, and some members of her tribe had to dive down and sever the creature's arms before the corpse could be recovered.

In April 1935, a large Pacific octopus attacked a fisherman wading in waist-deep water at the entrance to San Francisco Bay. Another fisherman ran to the rescue, and after a formidable struggle, the creature was finally dispatched with a knife-blade between the eyes. It had a radial span of 4.5m (15ft).

And octopuses do not take kindly to being hooked. Mabel Marriot, writing in *Of Sea and Shore,* tells of some friends who had a nasty experience with a Pacific octopus that had a body the size of a man's head and very long arms. They were fishing off the Alaskan coast and had hooked the octopus. They tried to get it off the line with a gaff, but it went into all kinds of contortions. As it came nearer to the boat, it grabbed the gunwale and one of the people, wrapping its arms around his arms and legs. The man was extricated from the tangle with considerable difficulty and sucker marks were left on his skin for some time afterwards.

Another Hollywood employee, John Craig, film cameraman and author of *Danger Is My Business,* similarly learned a healthy respect for the Pacific octopus. He also enjoyed diving – the old-fashioned kind with a suit, brass helmet, and lead boots. One day, he was lowered to the seabed at San Benito Islands, off Baja California, and he found a large hole in the rocks. He entered the vertical shaft and at the bottom he saw two large octopuses. Instead of back-tracking, he remembered the advice given to him by Japanese divers. They had said that you should not make any rapid movements that might attract an octopus's attention. Rather, they suggested, you should stay quite still. The octopus might examine you with its tentacle but will ignore you if you remain motionless.

Craig stayed absolutely still and, sure enough, the octopuses behaved as the Japanese had said. One raised its tentacle, examined the outside of his diving suit, and then seemed to lose interest. Then he made his move. He slipped off his lead boots, inflated his diving suit, and began to float to the surface. But the move was premature. The larger of the two octopuses stretched out an arm and grabbed his leg. Fortunately, the octopus was sitting on gravel and had no hold on the rocks. It was jerked out of the hole and hauled to the surface still attached to Craig.

On reaching the surface, the diving attendants hoisted Craig out of the water and saw that the octopus had smothered him completely. They tried to tear it off and went about attacking it with axes. They kept one of the severed arms. It was 2.4m (8ft) long. This meant that the octopus was about 5.5m (18ft) across. If it had been on rock rather than shingle the octopus would have killed the diver.

In the Mediterranean, the common octopus – similar to one of the species we find around the British coast except that the Mediterranean ones grow to a larger size – has been known to take offence at any interference. Scuba-diver David Oldale knows that, for it could have cost one of his friends his life. Writing in the specialist magazine *Diver,* he described how he was diving off Cape Yeronisou, on the island of Cyprus, with Mark Caney of Cydive at Kato Paphos, and wanted to get some good pictures of large octopuses; but what he got was more than he bargained for.

Oldale and Caney spotted a suitable candidate and went to investigate. The animal squirted out a cloud of ink and, using jet propulsion, headed for the safety of some rocks. Caney grabbed the tip of an arm in an attempt to stop it disappearing. The octopus got free but instead of swimming away, it turned bright red – the usual hue for an angry octopus – formed itself into what is known as the 'umbrella or flower position', with arms extended, mantle wide open, and the beak fully exposed. The octopus, estimated to be well over 2m (6ft) across, enveloped Caney.

The diver moved slowly to shallower water, attempting to loosen the octopus's hold. At one stage it tore off his face-mask, but Caney was able to replace it. The two struggled on and eventually the octopus let go and hid among some rocks. At the surface, the diver discovered he had large circular purple and yellow bruises all over his arms and wrists. It was an illustration of how, in similar circumstances, a less experienced diver could have panicked and drowned.

Sir Grenville Temple wote about such an event in his *Excursions in the Mediterranean Sea.* A Sardinian sea captain was bathing in just 1.2m (4ft) of water when he was seized by an octopus and held under. When his body was discovered the octopus had entwined its arms around his limbs.

An even more sinister story is told in Clemens Laming's book *The French in Algiers.*

> The soldiers were in the habit of bathing in the sea every evening, and from time to time several of them disappeared – no one knew how. Bathing was, in consequence, strictly forbidden; in spite of which several men went into the water one evening. Suddenly one of them screamed for help, and when several others rushed to his assistance they found an octopus had seized him by the leg by four of its arms whilst clung to the rock with the rest. The soldiers brought the 'monster' home with them, and out of revenge they boiled it alive and ate it.

In the *Genoa Gazette* of August 1867, there was news of a carter who went to bathe near the reef at San Andria and was seized by a large octopus. There were many other bathers in the sea who witnessed the event but none was courageous enough to go and help the struggling man and he was drowned.

Admiral Baillie Hamilton wrote to Henry Lee, naturalist of the Brighton Aquarium in the late 1800s, telling of a particularly large specimen that lived in Gibraltar Bay. It had grabbed and drowned one of the soldiers who guarded the Rock.

Stories of monstrous octopuses in the Mediterranean have dropped off in more recent years. Marine pollution and tourist disturbance in one of the world's most polluted seas have made conditions intolerable. Maybe, with the introduction of the UN Regional Seas Programme and the clean-up generally of the Mediterranean Sea, the larger specimens will be found once more. After all, Mediterranean octopus stories have a history going back a long way.

Pliny the elder, a Roman scholar who lived from AD 23 to 79, wrote in *Historia Naturalis* about a kleptomaniac giant octopus. When Pliny was appointed Governor over a section of Spain around Granada in AD 73, he heard about an enormous 'polypus' that would come out of the sea each night at Carteia – an old Roman colony near Gibraltar – and raid the curing sheds. The creature would carry off salted tunny and devour them at leisure in its underwater hollow somewhere along the coast. Eventually, the local fishermen decided the stealing must stop and killed it. The body, so it was said, was the size of a barrel. It weighed over 318kg (700lb), and the arms were 9m (30ft) long.

Salted fish, it seems, was a delicacy enjoyed by many a giant octopus. Aelian mentions in his *De Natura Animalium* another incident in Italy in which a large octopus actually squeezed and broke open a barrel of salt fish in order to get at the contents.

Elsewhere in the world, cephalopods have featured in the diaries of explorers, naturalists and members of the armed services. Sperm whale expert T Beale, for example, was once attacked by a strange cephalopod while collecting sea-shells on the seashore of the Bonin Islands (Ogasawara-Gunto) in the western Pacific.

> I was much astonished at seeing at my feet a most extraordinary-looking animal, crawling toward the surf . . . It was creeping on its eight legs, which, from their soft and flexible nature, bent considerably under the weight of its body, so that it was lifted by the efforts of its tentacula only, a small distance from the rocks. It appeared much alarmed at seeing me, and made every effort to escape.

Beale was not too keen to catch the creature but nevertheless put his foot on the end of one of its arms. Several times, however, it was able to pull itself free. Beale eventually bent down to hold on with his hand and the two began a tug-of-war that he felt would pull the limb apart. Then, he gave it a sharp jerk.

> But a moment after, the apparently enraged animal lifted its head, with its large eyes projecting from the middle of its body, and letting go its hold of the rocks, suddenly sprang upon my arm, which I had previously bared to my shoulder, for the purpose of thrusting it into holes in the rocks to discover shells, and clung with its suckers to it with great power, endeavouring to get its beak, which I could see between the roots of its arms, in a position to bite. A sensation of horror pervaded my whole frame when I found this monstrous animal had affixed so firmly upon my arm. Its cold slimy grasp was extremely sickening, and I immediately called aloud to the captain, who was also searching for shells at some distance, to come and release me from my disgusting assailant.

They managed to kill the creature, which was about 1.2m (4ft) across, with a knife, and later Beale was able to identify it as a 'rock squid', a name given to it by whalers. But was Beale's encounter with a squid or an octopus? The behaviour is more reminiscent of an octopus – creeping about on rocks, fixing on with its suckers, and showing the ability to pounce.

The Reverend William Wyatt Gill, a missionary in the Hervey Islands in the South Seas, described in the *Leisure Hour* how, in the early 1870s, he discovered another mode of attack by octopuses.

A servant of his went diving for *poulpes* (octopuses), leaving his son to look after the canoe. After a while he appeared at the surface with an octopus covering his entire face. If his son had not torn the octopus loose the man would have suffocated.

Swedish pearl diver, Victor Berge, wrote in *Danger Is My Life* about a particularly ugly experience he had with a large octopus in the Straits of Macassar, between Borneo and the Celebes. Dressed in a heavy diving suit, he was lowered down to about 36.6m (120ft) and was about to pick up a shell when he felt a tapping on his arm. Berge instinctively pulled out his knife and slashed out. As he turned, he saw he had cut off two arms of an octopus. The other six quickly grabbed his ankles and tugged his legs so hard his head kept banging against the sides of his helmet.

Berge, by this time, realized he was in grave danger and tried to give the divers' SOS signal – four jerks on his safety line – to the attendant, a Polynesian diver called Ro, who was in the boat. But just as he stretched out for the line the octopus dragged him away.

> I knew I was going. Just before a wave of unconsciousness swept over me I threw up my arms, caught both lines, and gave four frantic pulls. There was an instant when I had the sensation of being pulled in two. Then I knew nothing.

In the meantime, Ro and two other crew members on the boat realized something was wrong and tried to pull Berge up for all they were worth; but he wouldn't budge. Ro then had a brainwave: he took several turns on the line and put them round a stanchion. As the boat lifted in the swell it should tug Berge free of whatever obstruction was holding him down. They had not, of course, seen that he was struggling with an octopus. On the next uplift it worked and Berge was hauled to the surface. He was not alone. The octopus, still attached to Berge's leg, was coming too.

Ro spotted the creature as Berge neared the surface. He dived into the water, fastened another rope around his almost unconscious friend, and returned to the boat for a knife. He jumped in again and began to cut away the octopus's arms. Berge was hauled free and pulled on to the deck. He was battered and bruised but still alive.

In *Appleton's American Journal of Science and Art,* a correspondent writes about an octopus that attacked a diver working on the wreck of a sunken steamer lying on the ocean floor off the Florida coast. The diver, described as a 'burly Irishman', was helpless under the water and had to be hauled to the surface where he promptly fainted. His companions on the diving raft were unable to tear the beast away until they gave it a sharp blow on the body.

A similar thing happened to a diver in the estuary of the River Moyne, in Victoria, Australia, but he was able to hit the octopus with an iron bar and bring it to the shore. The radial span was about 2m (6ft). And, in 1960, near Cape Town at the tip of South Africa, an oyster fisherman had a fight with an octopus which spared his life but made off with his gold watch!

South Africa – East London to be precise – was the site of another octopus incident many years earlier in 1856, when Major Newsome of the Royal Engineers went for his morning bathe in one of the deep rock pools left by the receding tide.

> One morning I took a header into one of these pools which was, perhaps, 20 feet long, 7 to 8 feet wide, and deep in the centre – 8 or 9 feet. As I swam from one end to the other I was horrified at feeling something around my ankle, and made for the side as speedily as I could. I thought at first it was only seaweed; but as I landed, and trod with my foot on the rock, my disgust was heightened at feeling a fleshy and slippery substance under me. I

> was, I confess, alarmed, and so, apparently, was the beast on whom I trod, and whom, I suspect, I thereby discomfited, as he quickly detached himself and made again for the water. Some fellow-bathers, whom I hailed, came to my assistance, and with a boat hook, on to which the brute clung, he was, eventually, safely landed. When extended he would have filled a hoop to five feet diameter . . . Had I not kept mid-channel, I believe it would have been a life-and-death struggle between myself and the beast on my ankle. In the open water, I was the best man; but near the bottom or sides, which I could not have reached with my arms, but which he could have reached with his, he would, certainly, have drowned me.

The western Pacific is another area where coastal peoples have stories of dangerous octopuses. The South Sea Islanders, for instance, are fearful of these creatures, and with, it seems, good reason.

In September 1984, two fishermen from the Pacific nation of Kiribati, who had been hunting octopuses with spears, were reported to have been held under water and drowned by several octopuses that were estimated to be 3.5m (12ft) across.

In the mid-1800s, Lawrence Oliphant wrote in his *China and Japan* about a Japanese exhibition similar to Madame Tussaud's waxworks only the figures were carved from wood. One display showed a group of women running away after having bathed in the sea. One of their number is shown to be caught in the folds of a giant octopus. Its eyes and mouth were made to move by a man hiding inside the model.

But has this representation of a really giant octopus any factual basis? In a Japanese book in three volumes called *Land and Sea Products* by Ki Kone, about Japanese fisheries and fish-curing, there are some matter-of-fact references to the giant octopus and some interesting illustrations. They were placed before Western eyes when an article, accompanied by pictures, appeared in the *Field* of 14 March 1874.

One picture shows a fisherman in a small boat being attacked by a giant octopus. The fisherman has reached forward from the stern and with a long-handled knife, rather like a whaler's flensing knife, has chopped off one of the arms of the creature. The picture does not seem to show any exaggeration; the man, the boat and the knife all seem to be drawn to the same scale. The octopus is huge. A second picture is of a fish market in which are hanging two cephalopod arms. The onlookers appear astonished at the size of the limbs. And, in a further illustration a fisherman is depicted catching cephalopods. He tosses crabs into the water and spears the

Planet Earth Pictures – Carl Roessler

A large relative of this sea snake swimming in waters off the Philippines could be the source of many giant sea serpent stories.

Planet Earth Pictures – Howard Platt

Penguins in the Southern Ocean beware of the voracious leopard seal (*above*). Its long and slender body, and its serpentine head, make it a prime candidate for more southerly sea serpent sightings.

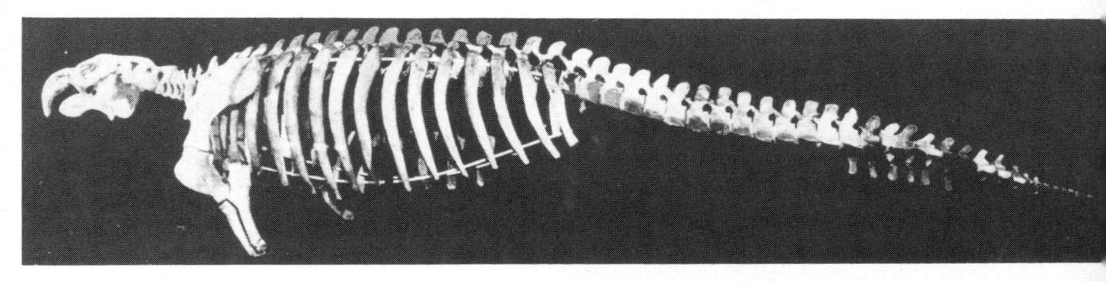

The Illustrated London News Picture Library

This is the skeleton of Steller's sea cow. It was dug from the peat on an island in the Bering Sea and bought by the British Museum (Natural History) in 1885. The species was thought to have become extinct at the turn of the century but there are isolated sightings of creatures resembling sea cows still reported today.

unfortunate octopus when it rises to the surface to catch the bait. Some people have interpreted the pictures as giant squid fishing, but the animal with the fishing boat looks very much like a giant octopus. And, in the last picture, the fact that the fisherman is using crabs again suggests that he is fishing for octopuses – giant ones.

These extraordinary giants were in vogue in the late eighteenth century when many a seafaring tale was told of octopuses so enormous they could turn over sailing ships. Pierre Denys de Montfort, in particular, introduced the world to the notion of the colossal *poulpe,* which made Pliny's tunny-stealing cephalopod seem rather puny. In his *Histoire Naturelle générale et particularière des Mollusques,* Denys de Montfort showed an enormous *poulpe* throwing its arms across a three-masted sailing vessel, with some reaching the very top of the masts. The drawing was purported to be of a picture that hung in St Thomas's chapel in St Malo.

The ship depicted, so the story goes, was sailing off Angola, West Africa, when the enormous octopus rose out of the sea and tried to capsize it. The crew prayed to St Thomas for his help in overcoming the monster, and with axes and cutlasses they were able to chop off the beast's arms and sail safely away. When the ship docked at St Malo the sailors went straight to the chapel and offered thanksgiving. A little later they commissioned a picture to be made of the horrifying event and then had it hung in the chapel. Researchers, who have since looked at Denys de Montfort's drawing, feel that the original artist had exaggerated somewhat. Indeed, Bernard Heuvelmans points out that the base of the arms would have had to have been more than 3m (10ft) thick, not the kind of limb through which mere hatchets and swords might cut.

Denys de Montfort was a conchologist at the Museum of Natural History in Paris and he started to write two important and well-received works on sea-shells. Unfortunately, his writings on octopuses were too sensational, to the point of being inaccurate, and he was distrusted as a crank and a fraud. He claimed, for instance, that six men-of-war captured from the French by Admiral Rodney and the British Navy in 1782 in the West Indies had been hijacked by giant octopuses, capsized, and sunk. The truth, however, was that the ships had not come to any grief whatsoever but had been sailed to Jamaica for a refit.

Denys de Montfort's enthusiastic massaging of the truth eventually caught up with him and he was relegated to obscurity. He was found dead in a Paris street in 1821, a broken man. Wouldn't it be ironical, though, if his suggestion, that there are truly giant octopuses living in the world's oceans, turned out to have a grain of truth in it?

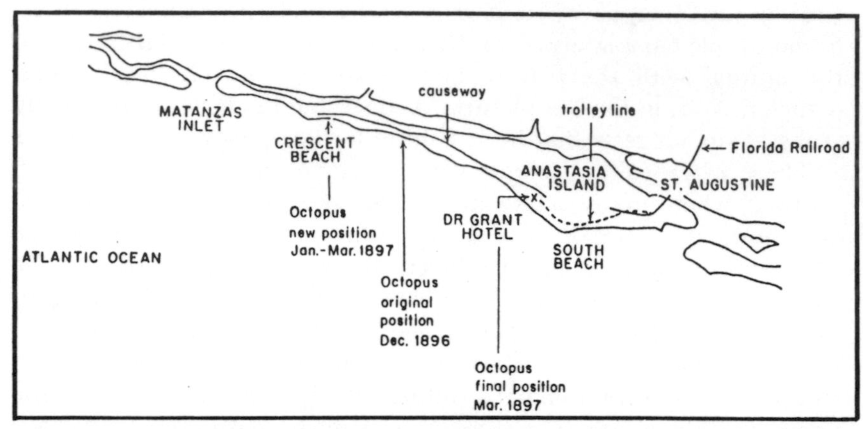

Map showing positions of the Florida monster carcass. *From a map by Roy Mackal.*

Modern science was not confronted with anything like a specimen until 30 November 1896 when two boys, cycling (so the story is told) along the seafront on Anastasia Island to the south of St Augustine, Florida, chanced upon the carcass of an enormous sea creature. It had been washed up on the tide and lay partly buried in the sand. Local GP and amateur naturalist Dr DeWitt Webb, President of the St Augustine Historical Society and Institute of Science, was summoned to examine the remains. At first he thought it to be a whale, but in a letter to J A Allen dated 8 December 1896 he provided a description of quite a different kind of beast – an octopus. The letter was forwarded by R P Whitfield to Yale University's Professor Addison Verrill, a distinguished expert on squid and octopus, and then published in the January 1897 *American Journal of Science.* Webb wrote:

> You may be interested to know of the body of an immense Octopus thrown ashore some miles south of this city. Nothing but the stump of the tentacles remains, as it had evidently been dead for some time before being washed ashore. As it is, however, the body measures 18 feet in length by 10 feet in breadth. Its immense size and condition will prevent all attempts at preservation. I thought its size might interest you, as I do not know of the record of one so large.
>
> The proportions given above indicate that this may have been a squid-like form, not an *Octopus.* The 'breadth' is evidently that of a softened and collapsed body, and would represent an actual maximum diameter in life of at least 7 feet, and a probable weight of 4 to 5 tons for the body and head. These dimensions are

> decidedly larger than those of any of the well-authenticated Newfoundland specimens (see Chapter 6). It is perhaps a species of *Architeuthis*. Professor Steenstrup recorded many years ago a species of this genus (*A. dux*), taken in 1855 in the West Indian seas, but his example was much smaller than the one here recorded.

The carcass, in the meantime, was on the move. Storms washed it out to sea and back again to Crescent Beach at the south end of Anastasia Island. And a little later Verrill published some additional information based on four photographs taken by Edgar Van Horn and Ernest Howatt and sent to him by Dr Webb. The photographs have since disappeared, but cryptozoologist Gary Mangiacopra uncovered some sketches made by Verrill's son, Hyatt Verrill, which give some idea of the shape and size of the creature on the Florida beach. Verrill wrote:

> These photographs show that it is an eight-armed cephalopod, and probably a true *Octopus*, of colossal size. Its body is pear-shaped, largest near the broadly posterior end. The head is scarcely recognizable, owing to mutilation and decay. Dr Webb writes that a few days after the photographs were taken (Dec. 7th), excavations were made in the sand and the stump of an arm was found, still attached, 36 feet long and 10 inches in diameter where it was broken off distally.
>
> This probably represents less than half of their original length, as the arms of *Octopus* generally taper very gradually and are often five or six times longer than the body. What looks like the remains of the stumps of arms is shown in the front view.
>
> The length, given as 18 feet, includes the mutilated head region. The photographs show that the 'breadth', given in the first account as 10 feet, applies to the more or less divergent stumps of the arms (?) and the body taken together, as they lie on the sand. The body, itself, is almost 7 feet wide, and rises at its thickest part 3½ feet above the sand in which it is partially embedded. The body is not greatly flattened, and probably had a diameter of at least 5 feet when living. The parts cast ashore probably weighed at least 6 to 7 tons, and this doubtless less than half its total mass when living.

Professor Verrill supposed that the creature was a species fed on by sperm whales and mentions stories told to him by whalers about the regurgitated remains of giant cephalopods. He went on to give the

Side view of the Florida monster carcass on a St Augustine beach.
From a drawing by Hyatt Verrill.

creature the scientific name *Octopus giganteus,* and proposed that it was related to the deep-sea, small-bodied, paddle-finned, jelly-fish-like octopod *Cirroteuthis.* He speculated that the posterior stumps noticed by Dr Webb were the remains of lateral fins, although he thought they might be a little too far forward. He went on:

> This is, at any rate, the first gigantic Octopod that has been described or figured from actual specimens.
> *Note.* Since the above was in type, I have learned that Dr Webb had caused the sand around the monster to be removed, and by means of six horses and powerful tackle he has moved the body higher up the beach. He says that the true length of the body is 21 feet. The head is mostly or entirely gone. The outer integument has dried to a firm mass several inches thick.

Dr Webb continued to supply additional information in letters both to Professor Verrill and to Professor William Healey Dall, curator of molluscs at the Smithsonian Institution in Washington DC, then known as the National Museum. Despite the excitement caused by the carcass, however, neither scientist, it seems, went to visit the site; and there the matter rested until a month later Professor Verrill did an about-face. In the *American Journal of Science* he wrote:

> Additional facts have been ascertained and specimens received, that render it quite certain that this remarkable structure is not the body of a cephalopod.

Verrill's identification of the beast as an octopus was based on the description supplied to him about the arms. It had been said that the stumps of arms had been 'adherent to one end' of the carcass, but further inquiries revealed this not to be the case. In a written statement made by a Mr Wilson (one of the first people who saw the carcass before it was washed along the shore and beached a second time) to Dr Webb it says of the arms:

> One arm was lying west of the body, 23 feet long; one stump of arm, west of the body, about 4 feet; three arms lying south of body and from appearances attached to same (although I did not dig quite to body, as it laid well down in the sand, and I was very tired), longest one measured over 32 feet, the other arms were 3 to 5 feet shorter.

Verrill was clearly right to be suspicious, for when the body was excavated no stumps or arms were found attached to the body at all. Could it be that the carcass was the remains of the head of a sperm whale and that the arms were those of squids which had been in the whale's stomach and which had been spilled out when the body had broken up on the beach? Verrill, I am sure, was thinking along those lines, and when Dr Webb sent him some pieces of the carcass, he was certain it had a closer affinity to whale than to octopus. Verrill wrote in the *American Journal of Science:*

> Dr Webb has recently sent to me several large masses of the integument of the creature, preserved fairly well in formalin. These masses are from 3 to 10 inches thick, and instead of being muscular, as had been thought, they have a structure similar to the hard, elastic variety of blubber-like integument found on the head of certain cetaceans, such as the sperm whale. They contain very little oil and cannot be called true blubber. They are firm, very tough and elastic, and composed mainly of much interlaced fibres and large bundles of tough, fibrous, white connective tissue. They are difficult to cut or tear apart, especially where indurated by partial drying. Some large irregular canals permeate the inner and less dense portions of the thick masses. These may have contained blood vessels originally. From the inner surface of some of the pieces large cords of elastic fibers proceeded inward. These now hang loosely from the masses of integument. Dr Webb states that these were found attached on all sides to a long saccular organ, which occupied most of the central cavity of the great mass. No muscular fibers were present in the specimens sent. Perhaps the muscular tissues of the inner surfaces, if any were present originally, have decayed, but the tough fibrous mass does not show much decomposition. The outer surface shows in some

places a tough, thin, gray, rather rough skin-like layer, that may be the remains of the outer skin. It looks a little like the skin of some fishes from which the scales have been removed. From these facts I am led to believe that the mass cast ashore is only a fragment, probably the head, of some huge vertebrate animal, covered with a blubber-like layer of great thickness.

Although such an integument might, perhaps, be supposed compatible with the structure of some unknown fish (the integument of the great sun-fish is very thick and elastic, but unlike this in structure) or reptile, it is certain that it is more like the integument found upon the upper part of the head of a sperm whale than anything else that I know. If we could imagine a sperm whale with the head prolonged far forward in the form of a great, blunt, saccular snout, freely projecting beyond the upper jaw, and with a great central cavity, it might, if detached and eroded by the surf, present an appearance something like the mass cast ashore. It hardly seems possible, however, that the abruptly truncated and narrow snout of the common sperm whale could take on, even after being long tossed about by the waves, a form like this. No whaler who has seen it has recognized it as part of a whale. It does not seem possible to identify such a large, hollow, pear-shaped sac, 21 feet long, with any part of an ordinary sperm whale unless its nose had become enlarged and distorted by disease, or possibly by extreme old age. No blowhole was discovered.

The specimen has now been moved several miles nearer to St Augustine and enclosed by a fence to protect it from the drifting sand. It is likely to remain in nearly its present state for several months more.

What happened eventually to the carcass is not told. Verrill, for his part, had written also to several learned journals, including *Science* – the journal of the American Association for the Advancement of Science, stating in the 5 February 1897 edition that his original identification of the carcass was wrong and that he was inclined to plump for a sperm whale as the identity of the beast. In the 19 March edition he confirmed this with more information about the specimens he had received.

In the same journal Professor Frederic Lucas of the National Museum was far more adamant about Verrill's revised view:

Professor Verrill would be justified in making a much more emphatic statement than that the structure of the masses of

> integument from the 'Florida monster' resembles blubber, and the creature was probably related to the whales. The substance looks like blubber, and smells like blubber and it *is* blubber, nothing more nor less. There would seem to be no better reason for supposing that it was in the form of a 'bag-like structure' than for supposing that stumps of arms were present. The imaginative eye of the average untrained observer can see much more than is visible to anyone else.

So, Verrill was firmly put in his place, and one wonders how much establishment pressure was put on him to change his original view. There again, the photographs do show a structure not unlike the head of a sperm whale and it does seem to be a plausible explanation. Verrill, however did point out there were problems with this interpretation and drew attention to the fact that several other scientists were supporters of the cephalopod theory.

There, then, the matter rested until 1957, when Forrest Wood, then director of Marine Studios (later Marineland) in Florida and now a scientist with the US Naval Research Center in San Diego, unearthed yellowing press cuttings from an old file and became intrigued by the 'Florida monster'.

On further investigation, Wood discovered from cephalopod expert Dr Gilbert Voss, of the University of Miami, that pieces of the creature had been taken to the Smithsonian Institution in Washington DC, where they had been wrapped in cheese-cloth and preserved in formaldehyde. Curator of molluscs, Harold Rehder, went to look for the 'Florida monster' material and a foray among the Museum's preserved specimen jars revealed that the specimens were still there, and labelled curiously '*Octopus giganteus Verrill*'.

Wood persuaded the authorities to allow a colleague, Dr Joseph Gennaro, then at the University of Florida and now at New York University, to take some of the material for tissue analysis. Gennaro visited the Museum himself and found the Florida material – 'half-a-dozen roast-size chunks' – in some murky liquid in a large glass jar. The tissue was very tough. Gennaro dulled four replaceable scalpel blades before he could cut away a couple of finger-sized slices. What struck him immediately was that the tissue was not at all oily, like blubber, and it had no distinguishing features. Examination under the microscope showed that the pattern of fibres in the tissue was more akin to octopus material than to squid or whale. After a series of tests, Gennaro concluded that the 'Florida monster' was, indeed, an enormous octopus.

Drawings from micrographs of sections of tissue samples from (a) squid (b) octopus and (c) Florida carcass.

But the scientific community still refused to accept the creature's identity. More proof was needed, they said. So Genarro enlisted the help of biochemist and cryptozoologist Professor Roy Mackal, who, at the University of Chicago, received samples of fresh octopus and squid, dolphin and white whale, some preserved squid and 'Florida monster' material. Each was only identified with a number, so that Mackal was unaware of what he was to analyse.

Mackal carried out a series of tests which compared the amino acids contained in each sample. He found that one sample contained high amounts of the amino acids glycine and proline, and, taken with the rest of the amino acids present, this indicated that it consisted mainly of collagen, the main structural protein of animal connective tissues. It is a substance with great tensile strength but little flexibility. The sample, it was later revealed, was from the 'Florida monster'. Mackal suggests that a large octopus would require such a tissue to support its enormous body.

Forrest Wood, meanwhile, had also been active in the giant octopus story. Even before he had found the 'Florida monster' clippings which triggered off these most recent investigations, he had come across references to such creatures in the Florida-Bahamas region of the western Atlantic.

He discovered, for example, that in 1941 there had been concern about the possibility of German submarines hunting off the Florida coast, and so the US Navy had carried out an extensive depth-charging programme that, as one witness described it, 'rearranged the ecology of the entire sea-floor'. A look-out on one of the vessels following the route of the explosions reported seeing a large, brown, kelp-like mass floating at the surface near Fort Lauderdale. As the ship moved closer, the sailor saw that it was not seaweed: 'As it moved into view there was no doubt as to its identity. The coils of its arms were looped up like huge coils of manila rope.'

Wood had also heard about giant octopuses on his excursions for Marine Studios to the Bahamas. There, he found, the Bahamians had many stories to tell about giant scuttles or lusca that live in the deep-water channels mainly around Andros Island. When Wood asked his local guide about the size of these creatures, the fisherman pointed to a shed about 25 metres away. Local folklore, he learned, told of giant scuttles entering shallow water only when sick or dying (reminiscent of giant squid stories?), and they were dangerous if ever they attached themselves both to the boat and to the bottom at the same time.

In another story, an island official told Wood of his encounter with a giant scuttle when on a fishing trip with his father. They were off Andros Island in about 180m of water fishing for silk snappers when his father thought he had snagged the bottom. But then they discovered that the line could be slowly drawn up. As it came nearer to the surface, they peered through the crystal-clear water and could make out the shape of a gigantic octopus. Suddenly, the animal detached itself from the line and clung to the bottom of the boat. Then, much to the occupants' relief, it let go and returned to the depths.

The island folklore is also enhanced when one considers the famous 'blue-holes' of the Bahamas, great cave systems that were first carved out by fresh water when, during the Ice Ages, the area was land and then flooded thousands of years ago when the glaciers melted and the land was swamped by the rising sea. Some of these are hundreds of metres deep (a blue-hole off Belize is over a thousand metres) and only a few have been explored.

Local people approach the holes with caution. They will fish from boats but on no account will dive into the seemingly black waters for fear of the lusca which, so the legend goes, drag men and boats to the centre of the earth.

Some of the blue-holes still have underwater channels connecting the main cavern with either the open sea or fresh-water lakes (the inland equivalent of blue-holes). Others have had their entrances blocked and in the bottom the accumulation of organic debris from above is broken down producing the easily identifiable hydrogen sulphide (bad eggs smell). One, on Bimini, to the west of Andros, has its roof intact and is entered via a tunnel to the side of the main cavern.

Holes connected to fresh-water lagoons can be dangerous because the salt water and fresh water do not mix. The layers are not constant but vary with the state of the tide. It is easy to see how legends of men being sucked down into the 'blue-holes' might have arisen. On incoming tides sea water pours in via the many channels and on outgoing tides fresh water is sucked out of the lagoons; whirlpools and swells are the result.

'Holes' connected with the sea could well have strange sea

creatures lurking in them, but as yet no cave diver has encountered one. What they do find, however, are caves guarded by sand sharks and others containing fish that swim upside down! Fish generally orientate to light and the only light in some of the overhangs is reflected from the sand on the bottom.

So, given the evidence from the Mediterranean, South Pacific, the Bahamas and Florida, might Bermudan fisherman Sean Ingham's mystery creature really be a giant octopus, a Bermudan relative of the Bahamian lusca? It cannot be totally ruled out, for there are many more curious cephalopods to be found in the sea and many are to be seen around Bermudan waters. Indeed, Ingham himself may have unwittingly found another kind of giant.

Dr Brian Luckhurst, senior fisheries officer for Bermuda, recalled the day Ingham came in on the VHF radio:

> On hauling up one of the crab-traps from about 500 fathoms, he felt as if the pull on the trap was very strong. As the trap approached the surface he could see a large mass covering it. It appeared to him to be a very large cephalopod with what he described as tentacles wrapped around the trap.
>
> He tried to hook it but the gaff kept on pulling through the tissue, and ripping it so that parts fell off. When it was pulled just out of the water, it began to break up on the wire mesh. In the end he only managed to get a relatively small piece in a bucket. Everything else fell into the water.

Luckhurst photographed the 22kg (49lb) chunk and put it into the freezer. The pictures were sent to the Smithsonian but have proved insufficient for any identification. The jelly-like appearance could indicate any number of sea creatures. Giant colonial salps, such as *Pyrosoma,* up to two metres long have been found drifting in the ocean, but Ingham is sure that he had brought up another deep-sea octopus, which he estimated to be about 9m (30ft) across.

One contender looks more like a jellyfish than an octopus, but an octopus is what it is – the deep-sea octopus *Allapossus.* It resembles a large gelatinous umbrella, with short arms and big eyes. The gelatinous octopuses were once thought to be quite rare, for very few have been caught in the nets of marine researchers. But, when the Americans lost their nuclear warhead off Spain, they took numerous photographs of the sea floor and discovered these umbrellas all over the place.

They are known to grow as long as 2m (7ft), 1.6m (5ft) of which is body and the rest arms, but it is thought that there may be larger specimens living in the deep. It is quite possible that Sean Ingham's octopuses, shaped like huge umbrellas, could have resisted being towed through the water in the manner of a sea-anchor, and given the impression of being very large and powerful.

Small species, such as *Vitreledonella,* are found in mid-waters, while the finned varieties, such as *Cirroteuthis,* live on the bottom. *Allapossus* is usually found above shelf slopes, just the kind of place through which Sean Ingham's traps must pass as they are hauled to the surface. A trap filled with crabs would be irresistible to a hungry octopus.

Not too long ago, Dr Malcolm Clarke from the Marine Biological Association in Plymouth was sent photographs of what could be a very large gelatinous octopus that had been brought up from 183m (600ft) in fishing nets to the west of Ireland. Its most striking characteristic is the size of the eyes. They are 15cm (6in) across. The animal's identity still remains a mystery.

But perhaps some of the biggest mysteries of all have been reports from the South Pacific and Chile of some very strange creatures that can only be described as 'big blobs'.

In *Great World Mysteries,* by Eric Frank Russell, there is a description from a diver of a huge black mass that drifted up from the depths in the South Pacific and engulfed a large shark the man had been following. And, in *Myths and Superstitions*, Julio Vicuna Cifuentes tells of stories from Chilean fishermen about a creature they call 'the hide' - a large flat octopus which '. . . has the dimensions and appearance of a cowhide stretched out flat. Its edges are furnished with numberless eyes, and, in that part which seems to be its head, it has four more eyes of a larger size.' The creature, it is said, will take anybody who enters the water.

The descriptions are reminiscent of the gelatinous octopuses. Could it be that there are enormous ones, of which we have little knowledge, living in the depths of the sea? Might, then, the fishermen's tales of the 'lusca' and the 'hide' contain a grain of truth? The suggestion, if nothing else, is certainly creepy.

6
The Kraken

On Canada's west coast, in 1892, the Hudson's Bay Company were moving their trading-post from the Nass River to Port Simpson, near the Alaskan border, and employed many local Indians to assist the operation. On one occasion, about 150 Indians in 50 canoes were towing a 30m (100ft) wide log boom. It was floating along with the current when suddenly it stopped, and all the paddling in the world could not shift it. Slowly, and much to the surprise of the Indians, the boom began to tow the canoes back against the current. Eventually, after many hours of strong paddling, the party managed to struggle to the landing site and at dusk they were able to beach the raft at high tide. In the morning, at low tide, they discovered an enormous squid, larger than the raft itself, squashed underneath. One arm was reported to be more than 30m (100ft) long and it ended in a large hook. The suckers were described to be 'as big as basin plates, to saucer-size at the ends'.

This amazing story was retold to Paul LeBlonde and John Sibert, and was included in their report 'Observations of Large Unidentified Marine Creatures in British Columbia and Adjacent Waters'. In this case, the creature could be identified. It was clearly a giant squid, probably of the genus *Architeuthis,* the largest of the cephalopods – the animal group which includes octopuses and squids – and probably the largest known invertebrates on earth.

Giant squids, unlike the giant octopuses, are known creatures. They have a comparatively short body, known as the mantle, eight short arms, and two long tentacles with a 'club' of suckers and hooks on the end (the squid has ten appendages, whereas the octopus has only eight arms). They have exceptionally large round eyes, a hard horny parrot-like beak at the opening of the mouth, a long 'pen' (equivalent to the cuttlefish's cuttle) that gives some support to the mantle, and on the underside of the mantle is a directable siphon that can squirt water to provide the beast with a form of jet-propulsion, or squirt ink as a 'smoke-screen' for rapid escapes from predators such as sperm whales or large sharks.

The Canadian teller of the story above was Charles Dudoward and he had a second tale to tell. The 1892 incident was recalled by his grandfather, but in the winter of 1922 Mr Dudoward himself was to meet the creature. This time one was washed ashore near Mrs Roberson D Rudge's Port Simpson Hotel which stood on piles at the top of the beach near the high-tide mark. Twenty men, he recalled, were needed to drag the carcass across the beach until it rested in front of the hotel. The arms were estimated to be about 15.2m (50ft) long, the width of the building, and the one surviving tentacle was about 30m (100ft) long. Like the 1892 squid, the tentacle had a 25cm (10in) wide by 31cm (12in) long hook at the end. When the dead animal began to smell, the hotel management had it towed out to sea and dumped.

Giant squids, though, were not always so readily accepted as 'real' animals. For centuries, they were considered to be mythical beasts and given the same scientific status as sirens, unicorns and basilisks. The 'kraken', as the giant squid was known, was feared by mariners, and judging by the reports from the early chroniclers of marine life, the fear was understandable. Enormous size, power and ferocity were attributed to the kraken. It was capable, so the stories tell, of pulling a sailing ship over and drowning all the crew.

That last point might bring forth a wry smile, but don't let it broaden too far; hearken to the sobering tale of the *Brunswick* in the early 1930s. This true story is told by Commander Arne Grønningsœter of the Royal Norwegian Navy. He was captain of the 15,000 ton auxiliary tanker which was sailing between Hawaii and Samoa in the Pacific Ocean when it was 'attacked', on no fewer than three occasions, by giant squids. Each time, the events took place in broad daylight with Commander Grønningsœter just 15.2m (50ft) above the sea on the ship's bridge.

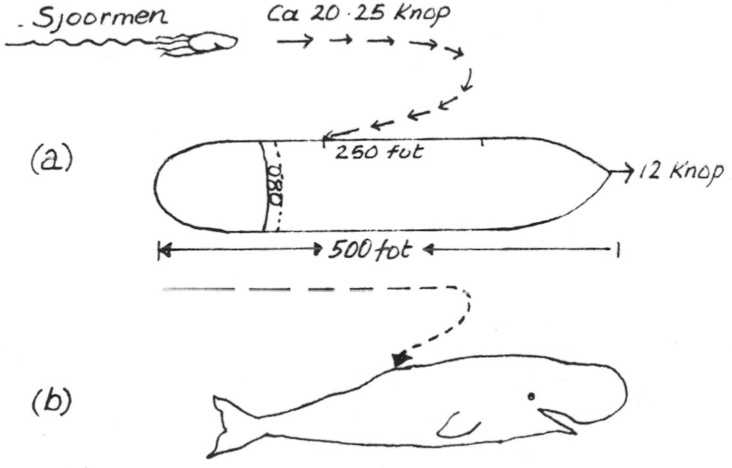

The attack by giant squid on the *Brunswick:* (a) the path taken by the squid; (b) compared to a sperm whale. *From drawings by Arne Grønningsœter.*

The squid, which were estimated to be about 9.1m (30ft) long excluding the long tentacles, came to the surface astern of the ship. They rapidly caught up, although the ship was travelling at about 12 knots, and when on a parallel course at a point 30m (100ft) from the bow, rapidly turned and slammed into the ship's side with a great thud. They then extended their tentacles, the thickness of 25cm (10in) pipes, in an attempt to fasten fast on to the ship. Failing to grasp the smooth steel hull, the squids slipped gradually towards the stern where they were chopped to pieces by the propeller blades.

Two aspects of the *Brunswick* story are interesting biologically: first, the squid deliberately attacked the ship, which they probably mistook for a whale; and second, the speed at which they overhauled the ship seems contrary to recent research findings.

The first point draws attention to the predator-prey relationship between giants squids and sperm whales. Traditionally, the sperm whale has been considered the predator. Yet, the *Brunswick* experience suggests that sometimes it is the giant squid that is the aggressor.

The significance of the second point only became apparent after some research in Norway in 1982. Until that time, all the squid that had been studied scientifically had been dead. On 23 August 1982, however, Norwegian fisherman Rune Ystebo caught a 10m long, 220kg giant squid in shallow water in a bay at Radoy near Bergen. It was still alive, but only just. Indeed, by the time Dr Ole Brix of the University of Bergen was able to take blood samples, the creature had been dead for two days.

Analysis of the blood was important to establish its oxygen-carrying capacity at different temperatures. There had been the feeling that the deaths of giant squids washed up or found floating at the surface in the recent and distant past were related to the temperature of the water; the warmer the water, the more giant squids found.

Dr Brix discovered that squid blood is not a good carrier of oxygen at the best of times, for example, at the cold temperatures in the depths of the sea. And at higher temperatures the carrying capacity is even worse. He concluded in an article in the scientific journal *Nature,* that giant squids are, in normal circumstances, sluggish swimmers. The research seemed to contradict the *Brunswick* experience: although giant squid might be capable of bursts of speed using their jet siphon to escape predators or attack prey, they are fairly sluggish for the rest of the time. He also suggested that they may suffocate when they experience raised ambient temperatures, a factor which might explain giant squid encounters in the past.

Most giant squid bodies have been found consistently at particular locations, for example, off Newfoundland or along the coasts of the British Isles and Scandinavia. These are places where warm water meets cold water. On the Grand Banks near Newfoundland, for example, the Gulf Stream pushes northwards against the south-flowing Labrador Current. A minor shift in the current could be a major problem for a giant squid. And along the Atlantic coasts of north-west Europe, the Gulf Stream brings large volumes of warm water at certain times of the year. A squid in the wrong place at the wrong time, i.e., in the warmer water, is likely to be a dead squid.

At the Marine Biological Association's laboratories in Plymouth, Dr Malcolm Clarke, suggests a further reason for unexpected giant squid appearances. The squid's buoyancy mechanism is also temperature-dependent. The relatively heavy body remains buoyant because high concentrations of ammonium ions (NH4), with a low specific gravity, are maintained in the muscles of the mantle, head and arms. In cold water the amount of ammonia is just right to maintain a neutral buoyancy, allowing the squid to go where it wants to go. In warmer water, the concentration rises and the squid becomes too buoyant. Consequently, it rises to the surface, cannot get back down again, and dies.

This is what is thought to have happened between 1870 and 1889 when the frequency of giant squid encounters off Newfoundland rose dramatically.

The first two individuals, recorded by the Reverend Mr Gabriel, were found washed ashore at Lamaline on the south coast of Newfoundland in the winter of 1870 – 1. They were 12.2m (40ft) and 14.3m (47ft) long respectively.

Then in October 1871, another dead giant squid body was found near the Grand Banks by Captain Campbell of the schooner *B D Hoskins,* of Gloucester, Massachusetts. He ordered a boat to be lowered to investigate and, when alongside, the mate measured the creature. It was a long, thin squid, the body being 4.6m (15ft) in length and 1.4m (4ft 8in) in circumference, with 3.1m (10ft) long arms that had suffered some damage. There is no mention of the tentacles.

An enormous 13.7m (45ft) squid was found at Bonavista Bay at the north end of Newfoundland in December 1872. The largest suckers on the arms were 6.3cm (2½in) in diameter. A 9.1m (30ft) specimen also came ashore there.

The following year, three Newfoundlanders had the shock of their lives when they came across a specimen that was well and truly alive. It was on 26 October 1873 that fishermen Theophilus Piccot and Daniel Squires, together with Piccot's son Tom, were out fishing for herring in a 6m (20ft) boat off Great Bell Island in

Conception Bay 14.5km (9 miles) from St John's. They spotted a large mass on the surface and, thinking it was wreckage, rowed over to take a closer look. Imagine their surprise when they poked it with a boat-hook and the mass reared up and shot out its two tentacles towards the boat; they had disturbed a kraken!

The creature grabbed the gunwale in its horny beak which the occupants described 'as big as a six-gallon keg', and wrapped a tentacle around the boat. The two fishermen were paralysed with fear, but not so young Tom. He seized an axe and chopped off one of the long tentacles and part of a shorter arm. Mercifully the squid sank back into the water, squirting defensive ink for all it was worth, and the fishermen rowed hastily for the shore at Portugal Cove, Tom still clutching his trophies.

On reaching their village, the short arm was thrown to the dogs, but the long tentacle was taken to a local naturalist – the Reverend Moses Harvey.

'I knew that I held in my hand the key of a great mystery,' said Harvey, 'and that a new chapter would now be added to Natural History.'

Before the events in Newfoundland, the kraken had been considered a mythical beast. Now it was a reality.

Harvey gave the squid tentacle to a representative of the Geological Commission of Canada who preserved it in alcohol. The length of the entire tentacle, before pieces had been chopped off, was thought to have been about 10.7m (35ft), and at its end was a 'club' shape about 0.8m (2ft 6in) long and 15cm (6in) in diameter. The overall length of the squid, from the tail end to the tip of the tentacles was estimated to be about 15.2m (50ft).

In November of the same year, another giant squid was caught in a herring net in Logie Bay, about 4.8km (3 miles) from St John's. It was also well and truly alive, and the fishermen had great difficulty extricating the beast from the net as it shot its tentacles through the mesh. Eventually, they cut the head behind the eyes so the body fell away into the sea.

Moses Harvey was quick to hear about it and offered the fishermen 10 dollars – a lot of money in those days – for the remains of the carcass. To allay their suspicions as to the value of the beast, he said he wanted to present it to the Queen.

They brought the body in a cart to his house and Harvey put it into a huge vat of brine in the hope of preserving it. He even dangled it over the side of his sponge bath to photograph it. Unfortunately, his attempts to preserve it intact were unsuccessful and he was forced to keep just parts of it. The animal was said to be about 9.8m (32ft) long, of which 2.4m (8ft) was the body portion.

And even larger squids featured in stories from the Labrador coast. They were reputed to have body lengths alone in excess of 24.4m (80ft), but these were unconfirmed reports.

In 1874 Professor Addison Verrill wrote in the *American Naturalist* about a specimen that had come close to rocks at Coomb's Cove in Fortune Bay and was hauled on to the shore by fishermen. Its reddish-coloured body, according to the Hon T R Bennett of English Harbour, Newfoundland, was 3.1m (10ft) in length and a long, thin tentacle was found to be about 12.8m (42ft). At its end was a clump of cup-like suckers with serrated edges.

In the same year, a 15.9m (52ft) specimen was washed up at West St Modeste in the Strait of Belle Isle on the Labrador coast, an unmeasured one was found at Harbour Grace, and a 12.2m (40ft) long one was stranded at Grand Bank in Fortune Bay.

Out at sea, Captain J W Collins, on board the schooner *Howard* reported seeing an extraordinary sight. The sea, he said, was full of dead and dying giant squids. Fishermen used the carcasses for fish bait.

During the next few years the squids continued to appear mysteriously: Hammer Cove in Notre Dame Bay, Lance Cove in Trinity Bay, Three Arms, Brigus in Conception Bay, and a stranding at James Cove. It was at Catalina Beach in Trinity Bay that another enormous giant squid was washed ashore after an equinoctial gale on 22 September 1877. It was alive when first found but quickly expired after exposure at low tide. Two fishermen took possession and they were advised by somebody in the large crowd of villagers who had turned out to witness the occurrence to take the carcass to St John's, where they would obtain a fair price for it. The Peabody Museum and the Smithsonian vied for the body but eventually the highest bidder was the New York Aquarium to which it was transported in a tank of methylated spirits. The official measurements were: body length 3.1m (10ft); tentacle length 9.1m (30ft); short arms length 3.4m (11ft); body circumference 2.1m (7ft); width of caudal fin 0.8m (2ft 9in); diameter of the largest sucker 2.5cm (1in); and it had 250 suckers on each of the short arms.

But the giant among giants turned up at Thimble Tickle on 2 November 1878.

Stephen Sperring and two colleagues were fishing not far from the shore and, like Theophilus and Tom Piccot and Daniel Squires, they spotted an object in the water they took to be some wreckage. They went closer to investigate and found the mass to be animate. Indeed, they had chanced upon the largest authenticated giant squid known to science.

The first thing they noticed was an eye. It was huge, about 0.5m (18in) across. Then they saw that the animal had run aground and was churning the water as it attempted to get back out to sea. It squirted large volumes of water through its siphon and flailed its arms about.

As the creature was partially disabled, the fishermen decided to try to catch it. They attached a strong rope to their grapnel-like anchor, which had barbed points, and threw it into the body. They took the rope to the shore and tied it to a tree so that when the tide went out, the squid was left high and dry. The fishermen kept a respectful distance as its long tentacles darted out from the central mass. Eventually it died and they chopped it up for dogs' meat. They did, however, have the good sense to measure the carcass. The body, from tail to beak, was 6.1m (20ft) long and the long tentacles 10.7m (35ft). The largest suckers on the ends of the tentacles were 10cm (4in) in diameter. Estimates put the weight at about 1.8 tonnes. It was truly a gigantic creature.

However, marine biologists feel that there are even larger ones to be found in the oceans of the world. In the Southern Ocean, for example, they think it likely that there are large shoals of these monsters patrolling the nutrient-rich waters of the Antarctic. They hunt at night. In days gone by, whalers used to slice open the stomachs of huge bull sperm whales and out would pour beaks and partially digested portions of long tentacles that could only have come from really gigantic squids. Abbé Pernetty, sailing with Bougainville in the Falklands in 1763 and 1764, knew about these, but nobody believed him when he returned with stories in his *Voyage aux Iles Malouines* of 'the coronet', which fastened on to ships with hooks, causing them to capsize, and was the biggest fish in the sea, according to South Seas sailors.

The largest specimen found in Antarctic waters, according to Malcolm Clarke, was a specimen of the gelatinous cranchid *Mesonychoteuthis hamiltoni* taken from the stomach of a sperm whale; it had a body length, excluding tentacles, of 3.5m (11ft 6in). Clarke says that other larger specimens were taken by the British research ship *Southern Harvester* during its cruise in the Bellingshausen Sea in 1955-6. One example, which was sent in formalin to the British Museum (Natural History) in London, has a body length of 3m (9ft 10in), and Dr Anna Bidder, of Cambridge University, has a section of a pen from a *Mesonychoteuthis* that suggests a body length of 5m (16ft 5in). And there might be even bigger ones.

A Soviet survey ship, in December 1964, had lowered instruments deep down in to the sea off the Ross Barrier when the heavy steel hawser was severed by something very large and very powerful. And, in 1968, the crew of a Soviet whale-spotting helicopter working in the Indian section of the Southern Ocean, saw what they thought was a body of a gigantic giant squid moving violently just below the surface. Closer inspection revealed it to be a brown colour and it had arms estimated to be a metre (39.4in) across.

In other parts of the world, whale fishermen have also come across gigantic squid that have been taken by sperm whales. On 4 July 1955 Robert Clark found the intact body of a 10.5m (34ft 5in) squid in a 14.3m (47ft) long sperm whale towed into the Fayal Island whaling station in the Azores. It weighed 184kg (405lb). Another 204.5kg (450lb) giant was found still alive in the stomach of a bull sperm whale caught by a Soviet whaler in the North Pacific on 31 December 1964.

In *The Cruise of the Cachalot*, Frank Bullen, who joined a whaling ship while in his teens, recalls the aftermath of a terrific struggle with a bull sperm whale in the Atlantic.

> During the conflict I had not noticed what now claimed attention – several great masses of white, semi-transparent-looking substance floating about, of huge size and irregular shape. But one of these curious lumps came floating by as we lay, tugged at by several fish, and I immediately asked the mate if he could tell me what it was and where it came from. He told me that, when dying, the cachalot always ejected the contents of his stomach, which were invariably composed of such masses as we saw before us; that he believed the stuff to be portions of big cuttle-fish, bitten off by the whale for the purpose of swallowing, but he wasn't sure . . . and sticking the boat-hook into the lump drew it alongside. It was at once evident that it was a massive fragment of cuttle-fish – tentacle or arm – as thick as a stout man's body, and with six or seven sucking discs or *acetabula* on it. These were as large as a saucer, and on their inner edge were thickly set with hooks or claws around the rim, sharp as needles, and almost the shape and size as those of a tiger.

Claws were evident in another specimen. In March 1769, Lieutenant Cook (later to become Captain) was sailing in HMS *Endeavour* in the North Pacific, latitude 38°44′S and longitude 110°33′W when a dead 2.1m (7ft) long giant squid, floating on the surface, was chanced upon. Joseph Banks, one of the two scientists on board, noted that the clubs at the end of the tentacles and the inner surfaces of the arms had, instead of suckers, a double row of very sharp retractable talons, much like those of a cat. The carcass was hauled on board and although somewhat mutilated by seabirds was put to good use in the form of calamary soup to vary the rather monotonous diet. Some parts, including hooks, were preserved and brought back to the Hunterian Museum of the Royal College of Surgeons.

More recently, hooks figured in another giant squid story. A US Navy submarine apparently bumped into a giant squid and the animal grappled with the vessel, mistaking it perhaps for a whale. It ripped the rubber cover off a piece of equipment on the hull. When the ship surfaced and the damage was inspected, the crew found a hook caught in the rubber housing. This hook, though, was the largest squid sucker-hook that had ever been found, suggesting another truly gigantic creature living in the depths.

The largest squid to figure in a mariner's tale, but which was not caught or accurately measured, was seen at night during the Second World War. The creature appeared alongside an Admiralty trawler lying off one of the Maldive Islands in the Indian Ocean. The witness was A G Starkey, who often fished at night over the stern of the ship using a cluster of light bulbs to attract the fish. One night, according to his story told in *Animals* magazine, he had an unusual visitor.

As I gazed, fascinated, a circle of green light glowed in my area of illumination. This green unwinking orb I suddenly realized was an eye. The surface of the water undulated with some strange disturbance. Gradually, I realized that I was gazing at almost point-blank range at a huge squid.

I say 'huge' – the word should be 'colossal', as so far all I could see was the body, and that alone filled my view as far as my sight could penetrate. I am not squeamish, but that cold, malevolent, unblinking eye seemed to be looking directly at me. I don't think I have ever seen anything so coldly hypnotic and intelligent before or since.

I took my quartermaster's torch and, shining it into the water I walked forward. I climbed the ladder of the fo'c'sle and shone the torch downwards. There, in the pool of light, were its tentacles . . . these were at least 24in thick. The suction discs could clearly be seen. The ends of the arms appeared to be twitching slightly, but this may have been a trick of the light.

My heart was going like a sledgehammer. Remember, I was alone on the deck, everyone else turned in. I was not so much afraid as excited, as if this were an opportunity to see something rarely seen by man.

I walked aft keeping the squid in view. This was not difficult as it was lying alongside the ship, quite still except for a pulsing movement. As I approached the stern where my bulb cluster was hanging, there was the body. Every detail was visible – the valve

> through which the creature appeared to breathe, and the parrot-like beak.
>
> Gradually the truth dawned: I had walked the length of the ship, 175ft plus. Here at the stern was the head or body and at the bows the tentacles were plainly visible . . . The giant lay, all its arms stretched alongside, gazing up, first with one then both eyes as it gently rolled. After 15 minutes it seemed to swell as its valve opened fully and without any visible effort it 'zoomed', if I may use the expression, into the night.

Mr Starkey's squid was truly enormous, but his story would almost be unbelievable if it were not for other similar tales of gigantic creatures. In 1926, for example, a badly damaged carcass which locals claimed to be a giant squid, was washed ashore at Port Shepstone on the Natal coast of South Africa. All its arms and tentacles were missing, but estimates, based on the size of the body alone, put its overall length, with outstretched tentacles, at about 30m (100ft). Some researchers believe this carcass was that of a whale, but another specimen found at Flower's Cove on the Newfoundland coast in 1934 was positively identified and measured. It was 22m (72ft) long. One found in the same area in 1882 was claimed to have a length of 26.9m (88ft).

The largest authenticated giant squid found since 1900 was caught by the crew of a US Coast Guard ship patrolling the Great Bahamas Bank near the Tongue of the Ocean – a deep undersea section of the Atlantic Ocean to the east of Andros Island – in 1966. It measured 14.3m (47ft).

A 9.5m (31ft 2in) long dead giant squid was discovered by two doctors who were fishing near Luanco to the west of Gijon on the north-west coast of Spain in July 1968, and an 8m (26ft 11in) specimen was caught near the Flemish Cap Bank at the eastern edge of the Grand Banks by the Portuguese fishing trawler *Elisabeth* on 4 July 1972. It can be seen in the Aquario Vasco da Gama in Lisbon.

Sucker marks on the sides and heads of sperm whales have been used to estimate squid sizes, with lengths in excess of 39.6m (130ft) claimed. Professor Verrill found that a 9.8m (32ft) long individual had its largest suckers about 3.2cm (1.25in) in diameter and a 15.9m (52ft) one had suckers 5cm (2in) across. What, though, do we make of the marks on a sperm whale caught in the Atlantic with 13cm (5in) diameter sucker marks and another mentioned by Ivan Sanderson with 0.5m (18in) diameter marks? Are they really from squid suckers, the battle scars resulting from the struggle of two giants, or are they marks caused by external parasites?

And when we talk of *Architeuthis* – the giant squid – it appears we talk about more than one species. Professor Verrill analysed the sizes and shapes of the Newfoundland specimens and surmised in 1878 that there were two species represented – one with a thin body and very long tentacles, and another with a stockier body and shorter tentacles. Since then, many species have been recognized worldwide – *A. harveyi* (after Moses Harvey) with the long tentacles and *A. princeps* with the long body are the giants of Newfoundland; *A. martensii* measuring 3.7m (12ft) overall was found in a Tokyo (then Yedo) fishmarket in 1873; *A. mouchezi,* another 3.7m (12ft) specimen was caught by fishermen in the Bay of Tateyama, not far from Tokyo; *A. sancti-pauli,* at 7m (23ft), was found on a beach on the northern side of St Paul in the Indian Ocean; *A. nawaji,* measuring 7.6m (25ft) of which only 1.2m (4ft) was the head, was caught in the Gulf of Gascoyne in the Bay of Biscay; *A. kirki* was represented by a 9.1m (30ft) long specimen that was found alive amongst rocks at Cape Campbell in New Zealand and *A. stockii* was found by three boys on the beach at Lyall Bay, in the Cook Strait.

The longest species, known from only one specimen, is *A. longimanus.* It was found in 1887 on the 'Big Beach' in Lyall Bay, New Zealand. It was measured at 18.9m (62ft). It had a short 2.4m (8ft) long body with enormous 16.5m (54ft) long tentacles.

Today, 19 species of giant squid are recognized (at one time almost every find was given a new name) but some teuthologists (those who study squid and octopuses) believe there are only three species of *Architeuthis*: *A. sancti-pauli* represents all the giant squid in the Southern Hemisphere, *A. japonica* covers all those in the North Pacific, and *A. dux* all those in the North Atlantic.

It was amazing, however, that Professor Verrill, in 1878, could make the scientifically acceptable proposal of naming even two species of giant squid. Only twenty-six years previously, the giant squid was not considered a 'known' animal. There had been plenty of stories and a few specimens, but it took the scientific community a long time to accept that there was such a thing.

References to giant squid, over six feet (2m) long, were made by that early man of science, Aristotle, who lived between 384 and 322 BC, but the legend of the kraken appears to have started in AD 1180 when the Norwegian King Sverre wrote about an enormous creature the size of an island. Olaus Magnus, the last Roman Catholic Archbishop of Uppsala in Sweden was next to refer to the kraken in 1555. He told of two people lighting a fire on the creature's back, after which it sank beneath their feet and they drowned. He retold the story in his *Historia de Gentibus Septentrionalibus* which he wrote and published in Rome. He retired there after Gustavus Vasa supported the Reformation in Sweden. In his great work he described the kraken as a 'monstrous fish' and in the English translation by J Streater in 1658 it is written:

> Their Forms are horrible, their Heads square, all set with prickles, and they have sharp and long Horns round about, like a Tree rooted up by the Roots: They are ten or twelve Cubits long, very black, and with huge eyes . . . one of these sea monsters will drown easily many great ships provided with many strong Mariners.

The 'monstrous fish' of Olaus Magnus in 1555.

At the turn of the seventeenth and eighteenth centuries, physician and naturalist Christian Francis Paullinus, who lived between 1643 and 1712, described a monstrous animal that rose out of the sea around the coasts of Troms and Finnmark (northern Norway), at the northern tip of Scandinavia, and which was of such a size, a regiment of soldiers could parade on its back. At about the same time, the Dane, Bartholinus, told the story of the Bishop of Midaros who put up an altar on what he thought was a big rock only to find that he had climbed on top of a 'sleeping' kraken. The monster was supposed to have allowed the good Bishop finish his service, letting him return to the land before sinking back into the sea.

Kraken folklore continued in Scandinavia with creatures described as being 2.4km (1.5mi) across and having mast-like arms. The Bishop of Bergen, Erik Ludvigsen Pontoppidan, wrote one of the most quoted works on kraken in his *The Natural History of Norway,* published in 1752 and 1753. He described them as 'floating islands' with 'horns which . . . stand up as high and as large as the masts of middle-siz'd vessels.' He warned:

> It seems these are the creature's arms, and, it is said, if they were to lay hold of the largest man-of-war, they would pull it down to the bottom. After this monster has been on the surface of the water a short time, it begins slowly to sink again, and then the

> danger is as great as before, because the motion of this sinking causes such a swell in the sea, and such an eddy or whirlpool, that it draws down everything with it.

Pontoppidan also mentions the stranding of a small kraken on the rocks on the north Norwegian coast. The story was told to him by the Reverend Mr Friis, Consistorial Assessor, Minister of Bodø in Nordland.

> In the year 1680, a Kraken (perhaps a young and foolish one) came into the water that runs between the rocks and cliffs in the parish of Alstaboug, though the general custom of that creature is to keep always several leagues from land, and therefore of course they must die there. It happened that its extended long arms or antennae, which this creature seems to use like the snail in turning about, caught hold of some trees standing near the water, which might easily have been torn up by the roots; but beside this, as it was found afterwards, he entangled himself in some openings or clefts in the rock, and therein stuck so fast, and hung so unfortunately, that he could not work himself out, but perished and putrefied on the spot.

He went on to tell of another incident that occurred near Fridrichstad, in the diocese of Aggerhuus.

> They say that two fishermen accidentally, and to their great surprise, fell into . . . a spot on the water . . . full of thick slime almost like a morass. [Pontoppidan mentions that the kraken saves up its excrement and voids it at specific times in order to lure fish to within arm's length.] They immediately strove to get out of this place, but they had not the time to turn quick enough to save themselves from one of the Kraken's horns, which crushed the head of the boat, so that it was with great difficulty they saved their lives on the wreck, though the weather was as calm as possible; for these monsters, like the sea snake, never appear at other times.

Carolus Linnaeus included the kraken as *Sepia microcosmos* in his classic taxonomic work *Systema Naturae* in 1735, but apparently got cold feet about the entry and it disappeared from subsequent editions. It therefore was not until just under a hundred years later, in 1847, when Professor Johan Japetus Steenstrup, a Danish zoologist, quoted from the diaries of an Icelandic naturalist, Sveinn Paulson, in a paper presented to a meeting of Scandinavian naturalists, that the giant squid was officially recognized. Paulson described a *Kolkrabbe,* as it was known in those parts, which was washed ashore at Arnarnaesvick. Its tentacles were 5.5m (18ft) long and the body, minus the head, about 6.4m (21ft) long and 2m (6ft) wide. Steenstrup announced to the meeting that he had christened the creature *Architeuthis monachus.* There the matter rested in relative obscurity until another specimen was washed ashore at Aelbaek Beach in Jutland, Denmark, in 1854, which Steenstrup considered a different species and gave the name *Architeuthis dux.* He had to base his description on the beak and its muscles, for the rest of the creature had been chopped up for fish bait. The beak was reported to have been 23cm (9in) long.

Although Scandinavian countries have been closely associated with the early tales of the kraken, there have been surprisingly few strandings. The largest, a 13m (42ft 8in) specimen, was found by fishermen near Tromsø, just north of the Arctic Circle on the Norwegian coast. Another, slightly smaller 11.3m (37ft) squid was washed up on a beach at Kyrksaeterøra in 1896, and in October 1964 a 9.1m (30ft) one came ashore at Ranheim on the Trondheim Fjord.

The most widely quoted squid encounter story, however, took place on 30 November 1861, when the French dispatch steamer *Alecton* was steaming about 193km (120mi) north-east of Tenerife, in the eastern Atlantic. A strange mass was spotted in the water and the captain recognized immediately that they had chanced upon the legendary kraken. He decided to capture it.

The crew fired at it with muskets, but to no avail. They threw harpoons but these did not stay in the soft flesh for long. There was a strong musky smell in the air, most likely from the discharged ink that discoloured the surrounding water. After three hours of firing and throwing, they managed to put a noose around the tail and hauled the beast to the side of the ship. As they pulled it from the water, however, its great weight caused the rope to cut through the body and most of it fell back into the water. The captain called off the hunt and they took the tail piece back to Monsieur Berthelot, the French Consul at Tenerife who, in turn, sent it on to Paris and the French Academy of Sciences. The creature, which was thought to have had a body length without tentacles of 5.5m (18ft) and was coloured red, was given the name *A. bouyeri* after the captain, Lieutenant Frederick-Marie Bouyer.

Earlier in the nineteenth century, another Frenchman, the zoologist François Peron, wrote in his *Voyage de Decouvertes aux Terres Australes* about a live animal that was spotted near Van Diemen's Land (Tasmania), not far from his ship *Le Geographe.* It was a giant squid the size of a barrel that appeared to be rolling on the surface. The arms were about 2.1m (7ft) long and 18cm (7in) in diameter and were flailing about like a collection of great snakes.

And a French expedition to the island of St Paul in the Indian Ocean to observe the transit of Venus, which occurred on 9 December 1894, discovered a 7m (23ft) long squid – 2.1m (7ft) of body and 4.9m (16ft) of tentacles – washed up on the north shore of the island by a tidal wave. The zoologist on board, Monsieur Velain, thought it to be a new species and the photographer of the expedition, Monsieur Cazin, was able to photograph it for verification.

Other voyagers also came across the 'giant calamary' as they were known. Jean Quoy and Joseph Gaimard, writing about the zoological events on their voyage around the world and published in the *Voyage de l'Uranie* in 1824, tell of a dead specimen in the middle of the Atlantic Ocean, not far from the Equator. They estimated it to be about 91kg (200lb) in weight, and they were able to secure some pieces which were deposited in the Museum of Natural History in Paris. And Captain Sander Rang is quoted in the *Manuel des Mollusques* as having seen a dark red squid with short arms and a body 'the size of a hogshead'.

Nearer to home, the giant squid has been an occasional visitor. Only a few specimens have been found around Britain – an *A. monachus* with 4.9m (16ft) long tentacles and a 2.1m (7ft) long mantle was found on the west coast of Shetland in 1860. The largest suckers, examined by Professor Allman, were three-quarters of an inch across; on 2 November 1917, a 6.1m (20ft) long *A. harveyi* was beached at Skateraw to the south-east of Dunbar, on the Scottish east coast; a 5.5m (18ft) *A. clarkei* turned up at Scarborough, Yorkshire, on 14 January 1933; the largest recorded giant squid in British waters, measuring 7.3m (24ft) overall, went aground at the head of Whale Firth, on Yell in the Shetland Islands on 2 October 1949, and on 30 November another 5.87m (19ft 3in) specimen was washed ashore at the Bay of Nigg at the head of the Cromarty Firth; on 1 February 1957 an Aberdeen trawler caught a 7.3m (23ft 11in) giant squid off Rattray Head between Peterhead and Fraserburgh; in 1971 a 6.7m (22ft) specimen was taken in Scottish waters; and in 1977 an immature female *Architeuthis* about 6.5m (21ft) long was washed ashore at North Berwick.

The largest squid to be taken off the British Isles was a magnificent 12.2m (40ft) long specimen captured to the north-west of Inishbofin off the coast of Connemara in the west of Ireland. The date was 26 April 1875.

The crew of a small curragh spotted a congregation of gulls above an object in the sea and went to investigate. Sergeant Thomas

O'Connor, of the Royal Irish Constabulary, reported the event in the *Zoologist.*

> They pulled out to it, believing it to be a wreck, but to their astonishment found it was an enormous cuttle-fish, lying perfectly still, as if basking on the surface of the water. Paddling with caution, they lopped off one of its arms. The animal immediately set out to sea, rushing through the water at a tremendous pace.
>
> The men gave chase, and, after a hard pull in their frail canvas craft, came up with it, five miles out in the open Atlantic, and severed another of its arms and the head . . . The body sank.

The fishermen deposited the recovered portion in the Dublin Museum.

On the west coast of Scotland, a smaller cousin of the Connemara giant, a 2.7m (9ft) squid, was washed ashore at Broadford Bay on the Isle of Skye. The date was 12 January 1952, the day after a substantial storm in the eastern Atlantic. In the calm of the morning, PC John Morrisey was walking on the beach when he came across an unusual 'fleshy-looking' object partly buried in the sand. He went to have a look and gave it a kick. Suddenly, a long tentacle whipped out and caught him by the leg. The only way the policeman could get away was by leaving his wellington boot behind. Later he returned to the creature with a pair of garden shears and killed it. Dr A C Stephen, of the Royal Scottish Museum in Edinburgh, identified it as *Stenoteuthis caroli.*

The *New York Times, The Times* and the Liverpool *Dispatch* carried another 'giant squid *v.* man story' in April 1921. The Cunard liner *Caronia* was in the Atlantic Ocean, heading west, when it was hit by a tremendous storm. Each time the ship ploughed into a wave, the water would cascade over the bows. After the ship had been hit by a particularly heavy wave, the ship's carpenter went to inspect the damage and on the fo'c'sle he found a 3.7m (12ft) long giant squid wedged between the winches. The animal shot out its long tentacles and grabbed him. The carpenter responded by hitting the squid with an iron bar. Several passengers ran to his assistance and after quite a struggle the squid was killed and the man released. The squid was preserved and brought back first to the Liverpool Museum, but was diverted later, so giant squid expert Dr David Hepple of the Royal Scottish Museum told me, to the American Museum of Natural History in New York. He also added that he thought the reports were somewhat exaggerated and the squid was nearer to 1.5m (5ft) in length. Nevertheless, it certainly gave that carpenter a bad turn!

Also in the Central Atlantic, during the Second World War, there was an even more macabre story of a giant squid encounter. On 25 March 1941, the British troopship *Britannia* was sunk by the German raider *Santa Cruz* about 1931km (1,200mi) west of Freetown. Some men survived and clung to driftwood and life-rafts. Lieutenant R E G Cox and eleven other men were supported by a very small raft; only their head and shoulders were above water. Sharks were the number one fear, but one night something far more horrific happened. A giant squid came up from the depths and wrapped its tentacle around one sailor and pulled him under. A little later Lieutenant Cox was seized on the leg, but the beast let go. Cox recalls incredible pain as the suckers were pulled off. The next day he noticed that large ulcers had appeared where the suckers had gripped, and when he was rescued the medical orderlies were constantly treating these wounds. He wrote to Frank Lane, author of *Kingdom of the Octopus*, in 1956 telling him that the red marks were still there, and Professor John Cloudesly-Thompson of Birkbeck College of the University of London was allowed to examine them. He was able to confirm that they were sucker scars likely to have been caused from an attack by a giant squid.

Off the coast of Peru, a relative of the giant squid is feared by local fishermen because of its ferocity – it is considered a demon. It is the Humboldt Current squid *Ommastrephes (Dosidicus) gigas,* which grows to 3.7m (12ft) long and hunts in large shoals at night.

Sports fishermen, who have found that the squid often strip their prize marlin or swordfish to the bone, have turned to catching the squid itself. It is a curious sport. The fishermen go out at night and look like Ku-Klux-Klan members. They wear a pillow-case over the head for protection. When the squid are brought to the surface, they not only shoot about at high speed, but also squirt a high pressure jet of water and ink which deluges the boat and the fishermen lining the rails.

Michael Lerner, benefactor of the Lerner Laboratory in Florida, once caught a specimen which, it was claimed, weighed 136kg (300lb) and had a body 3.1m (10ft) long and tentacles stretching to 10.7m (35ft). The eyes were reported to be 41cm (16in) across.

Many of those hooked fail to reach the surface intact. The other squid strip them alive. The steel traces on fishing lines may be bitten in two. The fate of any fishermen who fell over the side would not be hard to imagine. Some consider squid among the most dangerous animals in the sea.

Certainly Cuban shark fisherman beware of the embrace of the giant squid. Writing in his *Les Requins Se Péchent La Nuit,* François Poli discovered a healthy respect existed locally for these dangerous creatures. In the English translation Poli writes:

They talked of the gigantic octopuses of the Caribbean, measuring 50 feet across and capable of dragging down a 20-foot boat, or even seizing a man in a single tentacle and drowning him. These creatures never surfaced, they said, except on certain nights when the moon was full; then they floated for a few minutes, their phosphorescent eyes on a level with the water. These beasts moved with the speed of a shark, attacked everthing within reach of their tentacles and feared but one enemy – the cachalot. Cases were cited of captured whales whose bodies still bore traces of suckers the size of No Entry signs.

It is likely that in the translation, the word 'octopus' was substituted for the word 'squid', for the description clearly indicates that the Cuban fishermen were frightened of giant squid.

Giant squid, fortunately, seldom come into contact with man. If they did and the squid had food in mind, we would have little chance of survival, if some other eyewitness reports are to be believed.

Frank Bullen, in his *The Cruise of the Cachalot*, tells of another occasion in which the crew of the whaling ship were to witness the incredible strength of both the sperm whale and its adversary the giant squid. The ship had reached the Strait of Malacca, and after a long struggle with an enormous bull sperm whale during the day, the ship's crew had turned in, leaving Bullen on look-out duty.

At about eleven p.m. I was leaning over the lee rail, gazing steadily at the bright surface of the sea, where the intense radiance of the tropical moon made a broad path like a pavement of burnished silver. Eyes that saw not, mind only confusedly conscious of my surroundings, were mine; but suddenly I started to my feet with an exclamation, and stared with all my might at the strangest sight I ever saw. There was a violent commotion in the sea right where the moon's rays were concentrated, so great that, remembering our position, I was at first inclined to alarm all hands; for I had often heard of volcanic islands suddenly lifting their heads from the depths below, or disappearing in a moment, and, with Sumatra's chain of active volcanoes so near, I felt doubtful indeed of what was now happening. Getting the night-glasses out of the cabin scuttle, where they were always hung in readiness, I focused them on the troubled spot, perfectly satisfied by a short examination that neither volcano nor earthquake had anything to do with what was going on; yet so vast were the forces engaged that I might well have been excused for my first supposition. A very large sperm whale was locked in deadly conflict with a cuttle-fish, or squid, almost as large as himself,

> whose interminable tentacles seemed to enlace the whole of his great body. The head of the whale especially seemed a perfect network of writhing arms – naturally, I suppose, for it appeared as if the whale had the tail of the mollusc in his jaws, and, in a business-like, methodical way, was sawing through it. By the side of the black columnar head of the whale appeared the head of the great squid, as awful an object as one could well imagine even in a fevered dream. Judging as carefully as possible, I estimated it to be at least as large as one of our pipes, which contained three hundred and fifty gallons; but it may have been, and probably was, a good deal larger. The eyes were remarkable for their size and blackness, which, contrasted with the vivid whiteness of the head, made their appearance all the more striking. They were, at least, a foot in diameter, and, seen under such conditions, looked decidedly eerie and hobgoblin-like. All around the combatants were numerous sharks, like jackals round a lion, ready to share the feast, and apparently assisting in the destruction of the huge cephalopod. So the titanic struggle went on, in perfect silence as far as we were concerned, because, even had there been any noise, our distance from the scene of conflict would not have permitted us to hear it.

Bullen, thinking that the captain should be alerted, ran to his cabin and got very short shrift for disturbing him. On several occasions he wrote that he was surprised that the crew took such a little interest in events such as these, events that Bullen himself felt privileged to have witnessed.

The crew of a whale-catcher in the South Atlantic in the 1920s was certainly interested in the extraordinary event that they were fortunate enough to witness. It was described in Henry Bootes' book *Deep Sea Bubbles,* published in 1928. Bootes' own catcher-boat had just harpooned a female sperm whale and the crew were trying to get it back to the mother ship, the *Anna Lombard.* It was not an easy task. Sharks, many seabirds, and shoals of colourful fishes were trying to take their share of the disgorged pieces of squid, and there was the occasional monstrous visitor that took an unwelcome interest in the whale carcass.

The first unusual arrival was an enormous ray, about twenty feet across and coloured black with 'brilliant scarlet and deep blue spots', which the men speared, sending it to the bottom. The crew were fearful that these interlopers would damage the whale to such an extent that it would sink to the bottom of the sea and be lost as a prize. Then, what Bootes described as *Rorqualis australis,* arrived on the scene.

'Next to the killer,' Bootes said, 'this species of whale is pronounced the most ferocious.' To what species of whale he was referring, if it be whale at all, is not clear. He also gave it the name 'razorback' (a common name often given to the fin whale which is certainly not ferocious), and this particular specimen was about 10.7m (35ft) long, had a 'strange formation of its extenuated dorsal fin, and 'a massive head and jaws'. From the description, it is most likely that the 'killer' referred to a shark – perhaps the oceanic white tip which is the most likely to turn up at whale kills, and the 'razorback' was what we now call the killer whale or orca. Bootes wrote:

> This fin is a hard, very sharp bone formation with which it disembowels its victim, so said Casey, who had spent years in an American whaler in waters where these creatures were plentiful.

Continuing the story he described how it circled the carcass, much as a large shark would do, seemingly looking for a place to launch an attack, when, suddenly, all the small fishes disappeared. The reason for the exodus then became apparent.

> One particular spot of the impenetrable depths assumed a silver-whitish appearance, which at times became quite luminous, and very gradually we made out the waving arms of a giant cuttlefish. It gathered speed as it rose and I saw the awful eyes, which seemed to fix their gaze on me, holding me speechless and perfectly spellbound. I was quite unable to take my eyes off it, and my mind went back to Madam S. and our talk of this monster, on the terrace of her house in Valparaiso. The waving tentacles and long snake-like arms, each with rows of suckers, claimed my attention. As they waved upwards I could see them opening and shutting in anticipation of a feast. The body would be about twenty feet across the middle, but great portions of heaving flesh seemed to encase the joints, or sockets of the arms and tentacles, giving it greater massiveness. For ugliness, nothing that the morbid imagination of man has ever invented can compare with this pulsating horror. I speculated in my mind as to which would fall a victim to the tentacles – our whaleboat, the floating carcass or the razorback; and although these thoughts flashed through my brain, I made no attempt to order my men to pull out of the danger zone. We were all more or less hypnotized and helpless.
>
> Now I must say something of the razorback. We had made several attempts to beat it off, and each time it had dived under

> the carcass, tearing huge mouthfuls of blubber from the stomach, and evading our harpoon as though accustomed to such sport; but as the cuttlefish came closer, it too came under the spell of the waving arms, or was it the fixed gaze of the eyes and the rows of teeth (on the radulla?) which protruded from the strange-looking jaws, resembling in appearance a collection of parrots' beaks, or crab claws. Presently, with one mighty spring, it seized the razorback, not anywhere near the dorsal formation, but round the small of the tail and neck. Then the water became impregnated with the sepia which this vile thing ejects, and secure in the entwining embrace of the octopus the rorqual was carried to the depths below.
>
> When the water cleared all signs of the tragedy had vanished, but we still gazed into the silence, until the toot of the pinnace broke the spell.
>
> 'By the Holy M,' said the bowman nervously, 'that was a close call. Lucky for us the razorback came along or we might be making a trip to Davy Jones' locker, cuddled in the fond embrace of that slimy squid.'
>
> 'Those eyes!' said another with a shiver, 'I reckon I'll see them in my sleep until I die!'

In October 1966, two lighthouse-keepers at Danger Point, near Cape Town, were also privileged to see a rare sight. A whale, albeit a baby southern right whale, was on the receiving end of the wrath of a giant squid. For about an hour and a half the two men watched through binoculars as the squid clamped its tentacles around the whale calf's head and tried to drown it. As the calf gradually got weaker and weaker, its mother circled helplessly.

> The little whale could stay down for 10 to 12 minutes, then come up. It would just have enough time to spout – only two or three seconds – and then down again.
>
> We saw it lashing with its tail in the water and the squid's tentacles wrapped around its head. Each time the baby came up to breathe the mother whale surfaced with it, and then went down with her infant.
>
> The little whale was only just able to reach the surface and spout. The squid never moved. Finally we saw it come up for a second, blow, then go down. We never saw it again.

The crew of the Soviet whaler *Mirny,* fishing in the Antarctic in January 1965, also watched a tremendous battle between the traditional adversaries – a 40 tonne sperm whale and a 200kg giant

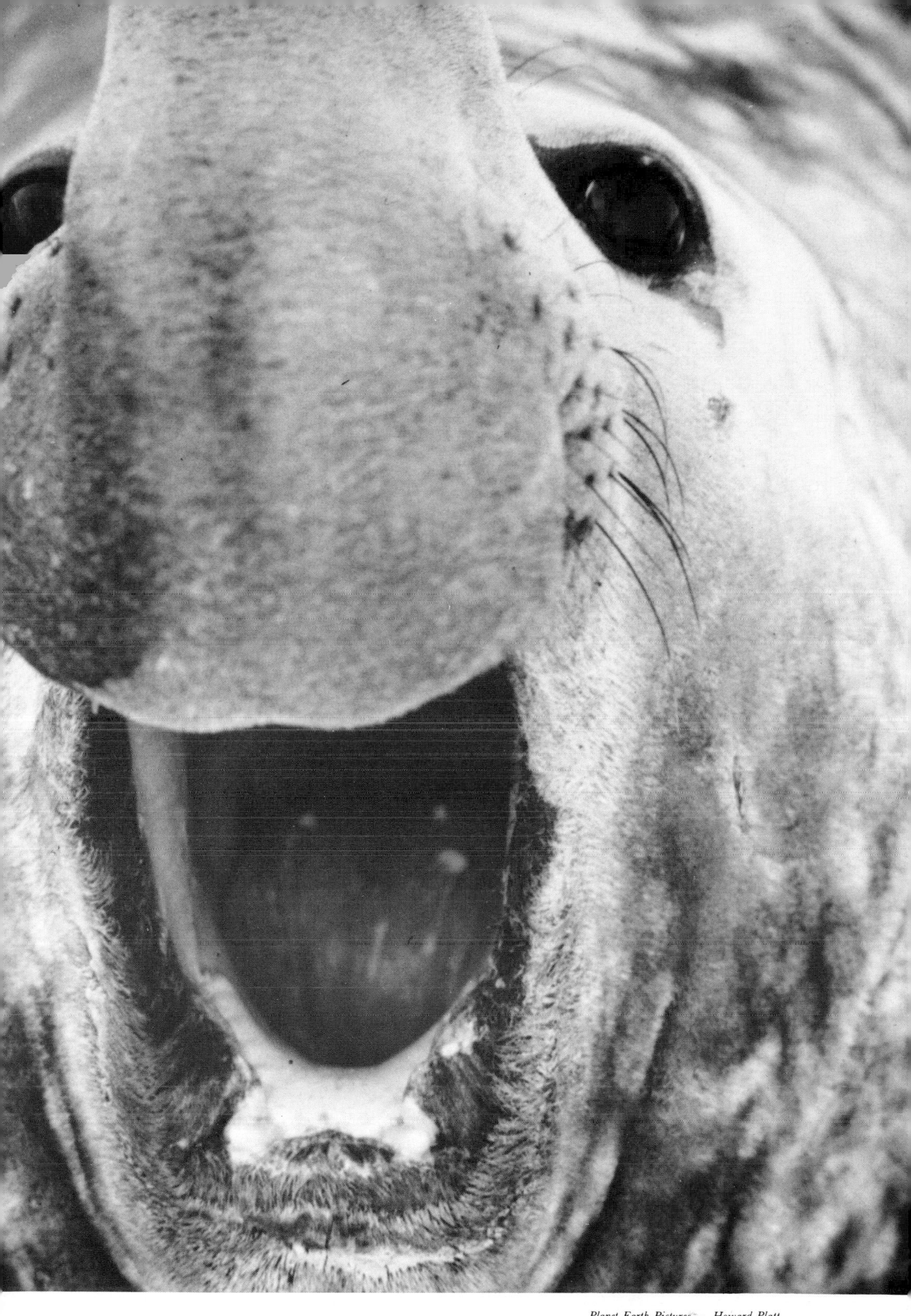

Planet Earth Pictures – Howard Platt

The bull elephant seal is enormous. It is recognised by the bladder-like nose which acts as a resonator to amplify the guttoral, snorting sounds made by males in the breeding season.

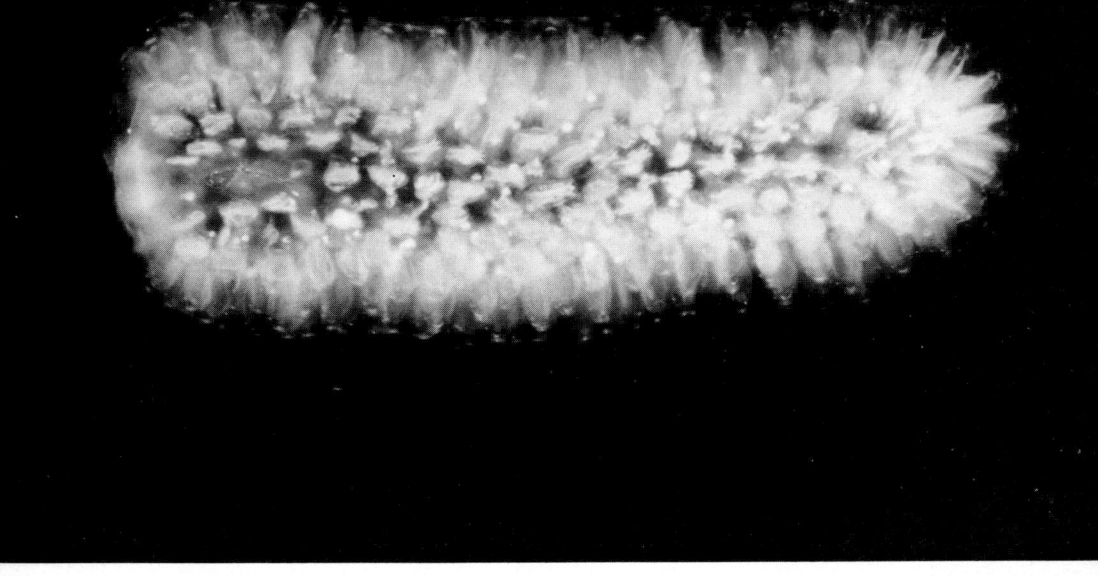

Planet Earth Pictures – Peter David

The salp is a short example of *Pyrosoma*. It consists of many individual organisms living in a single, luminescent, tubular colony. Some specimens have been found many metres in length.

This relatively small jellyfish has tentacles many metres long. Imagine the 'curtain of death' hanging below an Arctic jellyfish with a bell over two metres across. The tentacles would stretch 36 metres or more.

Jim Greenfield

Planet Earth Pictures – Richard Matthews

The anaconda from South American rivers can grow to immense sizes. Their aquatic nature makes them a prime candidate for 'Chessie'

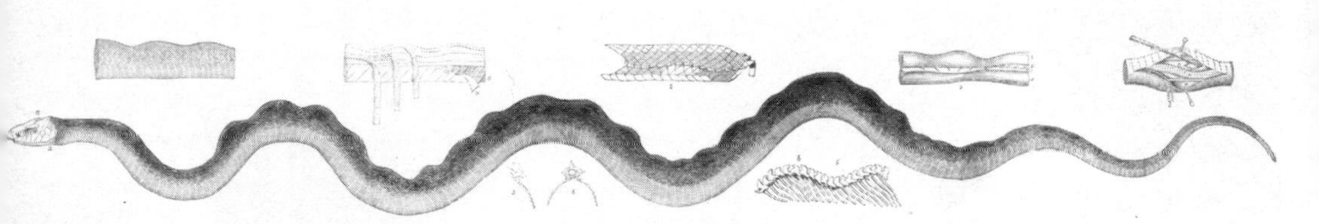

The Illustrated London News Picture Library

This drawing was labelled 'The Great American Sea Serpent' and given the scientific name *Scoliophis atlanticus*. It was thought to be a young sea serpent but turned out to be a common black snake.

The Associated Press Ltd

An albino saltwater crocodile was caught in the McArthur River in the Northern Territories of Australia. It was 5.5 metres long.

squid. The fight ended, according to reports, in the death of both giants. When the two corpses came to the surface the whalers harpooned them and dragged them on to their factory ship. They discovered that the whale had swallowed the squid's head, but that the squid had strangled the whale with its tentacles gripped tightly round its throat.

Gargantuan battles, such as these, between very large giant squid with 23cm (9in) long horny beaks and enormous bull sperm whales 15.2m (50ft) long with lower jaws armed with rows of 15cm (6in) long peg-like teeth, must be incredible sights to witness. But, how does a box-shaped whale catch a jet-propelled squid in the first place? The answer, it seems, is by sound.

Sperm whales communicate, in the main, by making clicking sounds. These can be heard quite clearly when following a herd of sperm whales in a sail boat. Occasionally, though, you might hear a very loud crack, almost an explosive sound. This, researchers believe, is a very high-intensity beam of sound which the sperm whale focuses through its spermaceti organ – a kind of 'sound-lens' – and directs at its prey. The unfortunate animal on the receiving end is debilitated and may even be killed, allowing the huge whale to catch up and swallow it with ease. The sperm whale's teeth are designed to grab and hold slippery prey rather than to cut up the flesh, so the whale slurps down the squid whole. The battles are likely to be the result of a sperm whale taking on a squid which is too big to be zapped into submission.

Sperm whale stomachs, though, are not usually filled with very large giant squid. Rather, they contain smaller squid and a host of other prey animals such as sharks, rays, and other large fish. Why, then, are the oceans not overrun with giant squids? Is there some other large predator in the depths that, together with the sperm whale, is controlling numbers?

Speculation aside – what do we know about the biology of giant squid?

Apart from Dr Brix's specimen found alive off the Norwegian coast, very few giant squid have been studied scientifically. The opportunity to do so, however, came in 1980 when a medium-sized squid, with an overall length of 10m (33ft), was washed ashore on Plum Island in Massachusetts. It was displayed for the public for a while and then taken to the Smithsonian Institution for detailed study by Dr Clyde Roper and Dr Kenneth Boss. First, they considered the anatomy.

The head, they revealed, is cylindrical and connected to the rest of the body by a short 'neck' surrounded by a collar. Eight thick arms and two slender but muscular tentacles surround the mouth which opens as chitinous, parrot-shaped jaws activated by powerful muscles.

Inside the mouth is a 10cm (4in) long rasping radulla, covered with chitinous teeth. It pushes the food, that has been chopped by the beak, into the first section of the alimentary tract, the muscular oesophagus. On reaching the stomach, enzymes are released which digest the food. Waste materials are expelled through the funnel.

The funnel is a modification of the molluscan foot. It joins the body just behind the head section on the underside of the animal as the first piece of the cone-shaped mantle or body. It points forward and, by pumping water out of the mantle cavity through the funnel, the squid can move backwards by jet-propulsion. The creature maintains stability with the aid of fins on the tail and rigidity with a pen, a skeletal support to which muscles are attached.

Giant squids tend to be dark purplish-red on the top side, and slightly lighter below. The skin contains both pigments and chromatophores. The latter are cells in which the pigment can be enlarged or contracted to facilitate visual communication, by changes in the pattern of the pigments in the skin – a kind of body-talk – with other squids.

Squids are primarily visual animals. They have the largest eyes in the animal kingdom, sometimes up to 25cm (10in) across. The lens is adjustable and there is a dark iris. It is thought the giant eye can be associated with an animal that lives in the deep ocean, although it is impossible to say at exactly what depth.

Evidence from giant squid fragments found in the stomachs of deep-sea sharks suggests *Architeuthis* is a bottom or near-bottom dweller in water that is about 1,000m deep. But a giant squid trawled up from 600m (1,969ft) in water that is over 4,000m (13,123ft) deep contradicts this assumption. And those squid found at the surface in Newfoundland in the 1870s must have been living in water only 100m (328ft) in depth.

Like Dr Brix, Drs Roper and Boss consider giant squids to be weak swimmers. Little is known about exactly what they eat for when squid are found they are usually 'sick' animals and their stomachs are empty. Researchers have guessed that they eat fish and other squid. Being sluggish, it is likely they lie in the water and grab fish or other squid as these swim by. They can shoot out the two long tentacles, grasp the prey with the club of suckers at the end, and haul it back to the beak where it can be held fast by the eight arms while the beak cuts it up.

The suckers are like suction cups rimmed with a chitinous ring with small teeth and can be moved slightly on their muscular base. The largest are on the wide section of the club, known as the manus.

Architeuthis species have no hooks in place of the suckers, but some other species do. The suckers on the arms are small at the tips and get larger towards the mouth.

Giant squid try to confuse their own predators with the aid of a 'smoke-screen'. Black ink is produced in a long gland that empties into the rectum and, when under attack, the animal can fill its mantle cavity with ink and pump it out into the sea via the funnel. Not only does the squid whoosh away, but it also leaves a dark mass of ink behind roughly the size and shape of its own body. The predator attacks the ink mass and the squid makes good its escape. Oppian, the third-century poet from Corycus (or Anazarbus), Cilicia, southern Anatolia, described it well in the Greek didactic poem 'Halieugica' on fish and fishing:

> The endangered cuttle thus evades his fears,
> And native hoards of fluids safely wears.
> A pitchy ink peculiar glands supply
> Whose shades the sharpest beam of light defy.
> Pursued, he bids the sable fountains flow,
> And, wrapt in clouds, eludes the impending foe.
> This fish retreats unseen, while self-born night
> With pious shade befriends her parent's flight.

Little, however, is known about reproduction. Male squids have two arms modified to grasp the female. Spermatophores – 10-20cm long packets of sperm that are transferred to the female – are found in ordered rows in an unusually long storage chamber, the Needham's sac. One end is extended into a penis-like structure shaped like a mushroom which protrudes from the mantle.

Each jelly-coated spermatophore (peculiar to giant squids), has an ejaculatory mechanism to release the sperm mass and a cement body to make sure it sticks fast when introduced into the female's mantle cavity.

The female produces large numbers of small white eggs. They are probably fertilized in the mantle cavity before being extruded into the ocean as transparent sausage-like strings surrounded by a protective layer of a gelatinous material. The jelly protects the eggs from fungal attack and from being consumed by fish. Inside the egg, so researchers from the Universities of Paris and Berne discovered, the embryonic squid is prevented from squirming too much and breaking out of the egg mass prematurely by a natural chemical tranquillizer.

Research on the giant squid, alas, has been on dead or dying animals. We have yet to see the creature living in its normal environment. But I believe it will not be long before we are able, perhaps, to witness one of the most remarkable confrontations in nature – the ultimate wildlife event – the battle between a giant squid and a sperm whale. Already, the National Geographic Society of the USA has been placing expeditions, equipped with deep-water cameras, off Bermuda and Newfoundland in the hope of filming giant squid. So far, they have seen six-gilled sharks – large primitively built sharks that live in the ocean depths – but the giant squid has evaded them. And a British commercial television company has been considering placing a remote camera system, triggered by the approach of deep-sea animals, on the floor of the Atlantic Ocean in the hope of seeing some of the strange and wonderful creatures that live there, including the giant squid.

On one occasion, underwater explorer and film-maker Jacques Yves Cousteau almost saw one. He was in a bathyscaphe when, through one of the portholes, he saw what looked like a white cloud. He switched off the vessel's lights and saw that the cloud was phosphorescent, just like the ink cloud of a disturbed cephalopod. Had he just missed seeing a giant squid in its natural habitat? On another occasion, he was luckier.

It was during one of his expeditions to the Indian Ocean in the research ship *Calypso* that the encounter took place. In his book, co-authored with Phillipe Diolé, *Octopus and Squid: The Soft Intelligence,* Cousteau recalls being in the minisub SP-350, a saucer-shaped submersible, and returning to the surface from about 300m (1,000ft).

> When I had reached 800 feet, I saw, through a porthole, a very large cephalopod, only a few yards from the minisub, watching the vehicle as it moved slowly past. I could not take my eyes from that mass of flesh, though it seemed not at all disturbed by the presence of the minisub. It was an unearthly sight, at once astonishing and terrifying. Was it sleeping? Or thinking? Or merely watching? I had no idea. It was there, none the less, enormous, alive, its huge eyes fixed on me. Then, suddenly, it was gone. I did not even see it move, though I am sure an animal of that size is able to move with extreme rapidity by means of water jets from the funnel. The impression it made was one of size and power. I can understand how formidable a giant squid must be.
>
> I wonder how many of us there are in the world who have ever seen such an animal: who have exchanged looks with those great, unblinking eyes, which resemble nothing I have ever seen before.

Cousteau, as far as I can ascertain, is the only person to have seen the giant squid in its normal domain, albeit very briefly. Let us hope that at the next encounter the cameras are rolling!

7
What Are They?

One March evening in 1933, an enormous 9m (30ft) high wall of water swept at great speed towards the Japanese coast. It engulfed the city of Sanriku, drowning 3,000 people, destroying 10,000 homes and sinking 8,000 boats. Survivors told how they saw strange lights in the waves and surmised that the wave was being pushed along by a gigantic luminous sea dragon.

In reality, the wave was a *tsunami* – a tidal wave – and was the result of an earthquake with a strength of 8.6 on the Richter scale. The quake had its epicentre many kilometres away somewhere below the Pacific Ocean. The magical lights were tiny phosphorescent organisms disturbed in the turbulance of the breaking wave.

The story illustrates how a perfectly normal, natural event (and quite a common event in that part of the world) is given a supernatural quality and embellished in the telling. It is as if we need to believe there are monsters in the sea. We need to be frightened.

Fear is a natural survival mechanism. Without it, our ape-like ancestors, descending from the trees a few million years ago, would have been caught and eaten by the plethora of predators waiting on the open grassy plains of the African savannah. Modern 'citified' man has little to fear from these creatures. The majority of people in the world today live in villages, towns and cities. We are isolated from the 'wild', rarely coming into immediate contact with dangerous animals. Yet, we still have the biological mechanisms to deal with such a confrontation and these are often activated when there is really nothing to fear. We actually simulate fear-inducing stimuli. The result is that we create images that might inspire fear – the mythological monsters of Greece, the dragons of China, and the horror movies of Hollywood – and these creatures have to be bigger and stronger than those in reality and have supernatural powers. Giant predatory cats have wings and can fly, snakes have a multiplicity of heads, and insects grow bigger than elephants. And, for a person floating gently on the surface of the ocean, peering down into its dark and mysterious depths, the imagination can run riot.

In the sea, a tree-trunk, floating at the surface with branches or roots sticking in the air, can look remarkably like a strange creature. The movement of the waves can almost bring it alive.

A few years ago, a small research vessel of the US Naval Oceanographic Office commanded by Captain F L Slattery, was conducting a hydrological survey of Chesapeake Bay – home of Chessie – when the crew's attention was drawn to a 'bright-eyed' sea monster. They saw the black silhouette of the long neck and head, the jaw flapping from the effort of swimming against the current, and went to investigate. What they discovered was a tree-trunk with a branch hanging like a jaw and a small hole through which the sun shone to form an 'eye'.

In the 1860s, there were two similar incidents. Ship's surgeon Arthur Adams reported in the *Zoologist* what he called an 'optical delusion'. At sunset, his ship was at the entrance of the Gulf of Pe-Chili in the Miatan group of islands. A light wind rippled the surface of the sea, and Adams watched as seabirds returned to their nests on the islands. Then, as he sipped on his 'congo', he saw a long dark object moving steadily through the water. He could see that it was not a seal or diving bird and so went on deck to enlist the help of others to identify the mystery creature. Many of the crew brought their binoculars to bear and concluded that it was a 3.7m (12ft) long snake moving against the tide and swimming with lateral undulations of the body. Adams wrote:

> So strong was the conviction that the course of the ship was altered, and a boat got ready for lowering. With a couple of loaded revolvers, some boat-hooks and a fathom or so lead-line, I made ready for the encounter, intending to range up alongside, shoot the reptile in the head, make him fast by a clove-hitch, and tow him aboard in triumph! By this time, however, a closer and more critical inspection had taken place, and the supposed sea monster had turned himself into a long dark root, gnarled and twisted, of a tree, secured to the moorings of a fishing net, with a strong tide passing it rapidly, and thus giving it an apparent life-like movement and serpentine aspect.

In *Nature,* Mr E H Pringle recalled a similar incident that befell the P & O liner SS *Rangoon* when sailing through the Strait of Malacca, bound for Singapore. The ship had just sighted Sumatra when an object was spotted off the port bow. The cry went up – 'The Sea Serpent'.

> And there it was, to the naked eye, a genuine serpent, speeding through the sea, with its head raised on a slender curved neck,

> now almost buried in the water, and anon reared just above the surface. There was the mane, and there were the well known undulating coils stretching yards behind.
>
> But for an opera glass, probably all our party on board the *Rangoon* would have been personal witness to the existence of a great sea serpent, but, alas for romance; one glance through the glasses and the reptile was resolved into a bamboo, root upwards, anchored in some manner to the bottom – a 'snag' in fact. Swayed up and down by the rapid current, a series of waves undulated beyond it, bearing in their crests dark coloured weeds or grass that had been caught by the bamboo stem.

Likewise, great floating mats of kelp can, to all intents and purposes, resemble a gigantic animal lying just below the water. Local water currents can mould the mass into a sinuous snake-like shape and give it a life of its own. Seen from a ship it could easily be mistaken for a sea monster. In December 1848, for example, the *Pekin* was becalmed at a position not far from the place where the crew of HMS *Daedulus* had seen their famous sea serpent a couple of months earlier (see page 107). On this second occasion a strange creature was spotted and examined through telescopes. It looked as if it had a huge head and neck and a shaggy mane. A boat was lowered and five men dispatched to investigate and capture the animal. The ship's captain Frederic Smith looked on with some concern for the safety of his men, but became curious when, looking from the main ship, he could see that the monster did not move when approached by the small boat. He then saw it being towed slowly behind. When they hauled the 'creature' aboard, they saw that it was covered in long-stalked goose barnacles, but underneath the animate covering was nothing more than a huge piece of seaweed about 6m (20ft) long and with a holdfast that resembled a head and neck. Most likely it had been ripped away during a storm and was fated to drift the ocean scaring the heck out of passing mariners.

A couple of months later, in February 1849, the *Brazilian* was stuck in the same area – latitude 26°S, longitude 8°E – when passengers and crew spotted what looked like a sea serpent. Thinking it to be the same object that was encountered by the *Daedulus,* the captain, Mr Harriman, decided to take a close look for himself. Armed with a harpoon, he stood in the bows of the small boat that had been lowered and was ready to tackle the beast. A passenger recorded the event:

> The combat, however, was not attended with the danger those on board apprehended; for on coming close to the object it was found

> to be nothing more than an immense piece of seaweed, evidently detached from a coral reef, and drifting with the current, which sets constantly to the westward in this latitude, and which, together with the swell left by the substance of the gale, gave a sinuous, snake-like motion.

The *Daedulus* affair itself has been explained away by L Sprague de Camp as the sighting of an abandoned native canoe being towed perhaps by a large marine creature such as a whale shark which the crew had harpooned. Maybe they panicked when the creature headed out to sea and they jumped out rather than cut the rope, or maybe they were tossed into the water when the shark surfaced under their boat. Hollowed-out log canoes or dugouts often have fierce faces painted on their bows and such a structure would look just like a rapidly moving sea serpent.

It is not always necessary for a tree, dugout or piece of seaweed to be present. Sometimes the surface movements of the water itself can give rise to sea monsters. A rip-tide, such as I have seen off the Needles and between the Welsh mainland and Skomer island, can create a wall of water that resembles the wake of a large animal swimming rapidly just below the water surface. The play of light and shadows on the disturbed water could suggest a sea monster to an untrained observer.

Also, groups of animals – dolphins or small whales in 'line astern' or a formation of sea birds, skimming the waves – can give the impression of a single large animal moving rapidly and undulating in and out of the water.

A young Henry Lee was walking along Brighton beach at twenty past eight on the morning of 16 February 1857 when his attention was drawn to the water by two boys shouting, 'A sea snake, a sea snake!' He wrote in the *Brighton Gazette*:

> I was induced to look in the direction of which they pointed. Coming from the westward, and about a quarter of a mile from the shore, I saw what I at first thought was a very long galley, very low in the water; but as it came towards and passed in front of us I saw it was that which the boys had pronounced it to be – a veritable sea monster. It was swimming on the surface, at the rate of from twenty-five to thirty miles an hour, and had exactly the appearance represented in one of the illustrated newspapers a few months since. I should say that about forty or fifty feet of it was visible, and I counted seven dorsal fins, if such they were, standing from its back. It continued in view for six or seven minutes.

Twenty-seven years later a much wiser Henry Lee (indeed the very same Henry Lee who had become a distinguished marine biologist) recalled his encounter on Brighton beach in this book *Sea Monsters Unmasked,* and came to quite a different conclusion as to the identity of his 'sea monster':

> I now know that the erect dorsal fins that I saw belonged to 'long-nosed porpoises' or dolphins, and, by their shape and height, am able to recognize their owners as having been of the species *Delphinus delphis.* My sea serpent was composed of seven of these cetaceans swimming in line, and, as is their wont, maintaining their relative positions so accurately that all the fins appeared to belong to one animal.

Similarly, at Gibraltar Point, on the north side of the Wash, a creature described by the *Skegness Standard* as 'The Thing' was sighted at the edge of the deep-water channel near the beach at the Point. A reporter from the newspaper found five eyewitnesses from Wainfleet who were on a day's outing to the coast, and they described the creature as 2.7 – 3.1m (9 – 10ft) long and dark in colour. Ray Handley, a butcher and ex-coastal defence operator, said that the thin black line they could see was about a mile or two away and travelling at high speed in a northerly direction along the edge of the deep water. He was puzzled that there was no wake.

A week later the *Standard* continued with the story, this time with more accounts from local people who had witnessed the passage of the mysterious creature. One group thought they had seen it submerge and re-emerge as it went along. Several mentioned the absence of a bow wave or wake.

Pauline North, a hairdresser from Skegness, supplied the answer to the mystery. She saw 'The Thing' heading north and thought it was a submarine until it reached a point in front of the Festival Centre skating rink and it took off. 'The Thing', it turned out, was a line of low-flying birds. A local naturalist filled in the details. The birds were scoters – gregarious sea-ducks often seen in huge rafts sitting on the sea's surface, but rarely seen close inshore.

It is not common to see birds in such numbers so close to the sea, but it is a phenomenon which has not gone unnoticed in the past. Writing in *Nature* in 1878, geologist Joseph Drew and several colleagues recalled a strange event while crossing the Channel from Folkestone to Boulogne. They saw a serpent-like object, 'about a furlong long', passing in front of the steamer. It was travelling at about 24 – 32kph (15 – 20mph). Fortunately, Drew had some opera glasses ('one of Baker's best') with him and was able to focus on the 'serpent'. He wrote:

> The first half of the monster was dark and glistening and the remainder of fainter hue, gradually fading towards the tail. The glasses did not determine the matter until the extreme end was reached, and then it was seen to consist of a mass of birds in rapid motion; those that were strong on the wing were able to keep up with the leaders, and to make the head appear thicker and darker by their numbers, whilst those that had not such power of flight were compelled to settle into places nearer and nearer the tail.

Drew identified the birds as shags, *Phalacrocorax aristotelis,* returning to their roosting sites after a day's fishing.

These observations, though, do not offer reasonable explanations for most sea monster sightings. In many instances, witnesses have almost certainly seen large animate objects, but here again, the mystery need not be that deep. There really are, after all, giants in the sea, and any of them could be mistaken for a sea monster.

Even the simplest of sea creatures can grow to immense size and under certain conditions, might be mistaken for something more unusual. Some species of jellyfish, for example, are very large. The Arctic giant jellyfish *Cyanea capillata arctica* is found in shallow waters in the North Atlantic. A specimen was found in Massachusetts Bay in 1865 that was 2.29m (7ft 6in) in diameter and had tentacles hanging 36.6m (120ft) below the bell. This net of death could cover an area of 457m (500yds) square. A related species or subspecies, the 'Lion's mane' jellyfish can grow to a metre across and has yellow tentacles up to 15m (49ft) long. Occasionally, 800m (½mi) wide swarms are seen drifting towards the Norwegian coast. The species prompted Sir Arthur Conan Doyle (himself a sea serpent witness) to employ its sting for criminal ends in the Sherlock Holmes book *Adventures With the Lion's Mane.*

When seen at the surface, the shadow of the pulsating bell of one of the larger specimens could give an observer the impression of seeing the tip of the back of a gigantic underwater creature – and in the case of the Arctic jellyfish it truly is enormous!

A group of 'watery' marine creatures known as the tunicates (more closely related to the 'higher' vertebrate animals than to invertebrates because the larvae possess a notochord which appears to be a precursor to a back-bone) are also contenders for sea serpent sightings. Many species are sessile and solitary, like the sea squirts or 'dead man's fingers' found in coastal waters, but some are colonial and float about in the surface waters of the open ocean – the salps.

A colonial salp *Pyrosoma* consists of an elongated cocoon-shaped tube, the walls of which are made up of thousands of individuals.

Each member of the colony plays host to symbiotic luminescent bacteria which gives the entire organism a glow which can vary from blue-green to yellow – hence the genus name meaning 'fire-body'. The luminescence is 'turned on' in response to touch, and it is told that the scientists on the *Challenger* expedition sometimes amused themselves at night by writing their names on the surface of colonial salp tubes. One species of *Pyrosoma* can grow to a length of up to 9m (30ft) and it is quite possible that the long, snake-like body could be mistaken for a sea serpent.

Worms grow to some surprising sizes, although what they gain in length they lack in width. The terrestrial giant earthworm *Microchaetus rappi* of South Africa can grow up to seven metres (23ft) with a body width of 20mm (0.79in). Worms have been seen straddling the six-metre wide highway near Debe Neck in the Eastern Cape. In Gippsland, Australia, another annelid, *Megascolides australis,* is found regularly with a length of over two metres (7ft). When held, and one end allowed to extend naturally, the creatures can reach a length of up to 4m (13ft). Might there be marine equivalents, perhaps, related to ragworms or lugworms? The largest known – the king ragworm *Nereis virens* – grows to 30cm (11in) in British waters, although a metre-long specimen was dug out of the mud on the Northumberland coast in 1975.

The monster marine worm idea to explain away aquatic monster sightings is not new. F W Holiday, writing in his book *The Great Orm of Loch Ness* in 1968, tried to give credibility to giant marine worms and proposed 'Nessie' to be a large version of *Tullimonstrum gregarium* – a 35.6cm (14in) long worm found only as a fossil and described by paleaontologists from the Chicago Field Museum of Natural History in 1966. Needless to say, few people took the theory too seriously.

The scientific community, though, were forced to take seriously a discovery made in 1977. One group of creatures found living next to the deep-sea hydrothermal vents on the Galapogos Rift (see page 9) consisted of three metre-long red vestimentiferan tube-worms *Riftia pachyptila.* They live in tough, flexible tubes and extend from one end as bright red plumes. The distinct colour is derived from the oxygenated haemoglobin (the iron-based compound that carries oxygen) in the blood. They have no gut, but play host to symbiotic bacteria from which they obtain nutrients. So, if sedentary tube-worms can grow to such a size, why not other more active marine worms?

Amongst the invertebrate animals, the most obvious contender for sea monster sightings must surely be the giant squid. It has size in its favour, and the variety of body parts – arms, tentacles, tail fins and so on – waving above the sea's surface could be mistaken for a writhing sea serpent. A long tentacle with its familiar club of suckers at the end might easily be taken for a long neck on top of which is discerned a small head – a classic sea serpent description. And a trail of discharged ink could resemble a snake-like body. The story of the attack on the *Brunswick* (see page 141) by a giant squid demonstrates this. In a letter to *Naturen,* the executive officer of the ship, Arne Grønningsœter, wrote:

> Since the wind at the time was fresh and the equatorial waters turbulent, I thought at first it was the giant sea serpent itself. But, when it started to attack, I realized it was a giant squid because of the inky black water it left behind. The animal moved very fast, and the trail of ink in the choppy sea really looked like a long snake.
>
> One of the reasons I could see so well was because I was standing on the bridge of this 15,000-tonner, about 50 feet above the water. I could imagine seafarers on sailing ships watching from a lower level and thinking that what they saw was a sea serpent.

The skipper of the steamer *Kiushiu Maru,* Captain Davison, and his chief officer were witness to an equally strange event just 15km (9mi) from Cape Sata-Misaki at the southern tip of Kyushu Island, Japan. At about 11.15 on the morning of April 1879, the two seamen watched as a whale leapt clear of the water. With the naked eye they could see what looked like something attached to it. By the time the whale jumped again they were able to see something holding on to the belly. Davison wrote of their experience:

> . . . myself and chief then observed what appeared to be a large creature of the snake species rear itself about thirty feet out of the water. It appeared to be the thickness of a junk's mast and after standing about ten seconds in an erect position, it descended into the water, the upper end going first.

Davison was sure they had watched a giant sea serpent battling with a whale, but was that the case? Could not the long snake-like creature in fact have been a tentacle of a very large giant squid? Those who have considered the sighting are drawn to the way in which the monster attached itself to the whale. A biting fish or snake would grab a flipper or the tail rather than latch on to the belly. A

The *Kiushiu Maru* sighting of 1879. *From a drawing by Captain Davison.*

clublike end of a squid's tentacle, however, would be best attached to a large, flatter surface like a whale's underside. It is now generally agreed that the battle Davison saw was between a whale and a squid.

A similar battle was observed a few years earlier by the crew of the barque *Pauline.* On 8 July 1875 the ship was at latitude 5° 13′S, longitude 35°W when it encountered three sperm whales. The crew saw that one of the whales had a serpent-like creature wrapped around its body. The whale and serpent whirled round and round for about fifteen minutes, and then the serpent, according to the crew's affidavit sworn before a Liverpool magistrate, dragged the whale to the bottom.

But which animal was doing the dragging? Is it not more likely that an enormous sperm whale – the largest known predator in the sea – would have the upper hand? The captain and crew of the *Pauline* say they saw a sea serpent. Bernard Heuvelmans suggests it was a giant eel. Frank Lane plumps for a giant squid. Who is right?

The *Pauline* sighting. *From an engraving.*

It seems to me that the most rational explanation is that the sailors on board the *Pauline* witnessed the gargantuan battle between a giant squid and a sperm whale. There is, after all, no record of a harpooned cachalot regurgitating portions of giant sea serpents or giant eels, but there are numerous references to giant squid.

Many cryptozoologists have proposed the giant squid as a rational explanation for sea monster sightings. Henry Lee, for example, interpreted Hans Egede's 'most dreadful monster' as the tail and arm of a truly gigantic squid.

The creature was seen on 6 July 1734 off Godthaab on the west Greenland coast by Egede, a Scandinavian priest known as the apostle of Greenland. He was, like many of the clerics at that time, interested in natural history and recorded what he saw in a book that was translated into English in 1745 and entitled *A Description of Greenland.* He recognized the presence of right whales, rorquals, killer whales and narwhals in the area, but this creature was of a different dimension altogether:

> This Monster was of so huge a size, that coming out of the Water, its Head reached as high as the Mast-Head; its Body was as bulky as the Ship, and three or four times as long. It had a long pointed Snout, and spouted like a Whale-Fish; great broad Paws, and the body seemed covered with Shell-work, its skin very rugged and uneven.

The sea serpent compared in size to a sailing ship as seen by Hans Egede and drawn by Pastor Bing in 1741.

One of Egede's colleagues drew a picture of the monster and it appeared as an illustration on a map in his son's (Povel Egede) book which continued the Greenland saga. What Henry Lee was able to do was to match the famous picture with one of his own, this time filling in the rest of the creature below the surface with the body of a giant squid. Some interpreters, such as R T Gould, rejected the notion. For a squid's tail to stick about 9m (30ft) out of the water (the average mast-height of ships at that time), it was felt that such a squid would have had to have a body length of at least 15m (50ft) and that squid of such a size just did not exist.

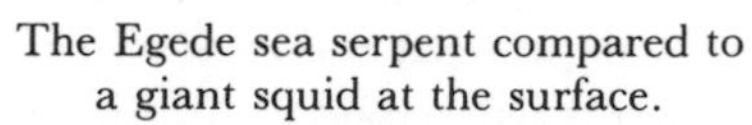

The Egede sea serpent compared to a giant squid at the surface.

I feel, though, we must question the original observations. Was the body really the same height as the ship's masts or were imagination and size-distortion factors, so common with maritime observations, operating here? Might the visible part of the body have been a little smaller? The Thimble Tickle squid (see page 145) was reputed to have a body length of 6m (20ft). That creature would have looked pretty impressive when viewed from the same angle and under the same circumstances as the Egede monster. And who is to say there are not larger squid, anyway? A creature of that size, with a buoyancy problem caused by a sea temperature change, flailing about at the surface and 'blowing like a whale' from its siphon, could easily be taken for a dangerous-looking sea monster. Interestingly, the sighting was also not so far from the area in which so many incapacitated squid had surfaced in the 1870s (see page 143).

When we consider the vertebrate animals, there is a whole galaxy of sea monster stand-ins. The cartilaginous fishes, including the sharks and rays, provide us with many suitable giants.

The largest fish in the sea is the whale shark *Rhiniodon typus*. It was not recognized by the scientific community until comparatively recently. In 1828, the first specimen caught was only 4.6m (15ft) long – the smallest whale shark ever found. A fisherman harpooned the fish in Table Bay, at the southern tip of South Africa, and a British Army surgeon, Andrew Smith, realized its importance, but it was not until forty years later that an Irish naturalist working in the Seychelles had specimens two and three delivered to him.

E Percival Wright had heard of a monster fish known locally as the 'Chagrin' (the French for shagreen – the material made from the skin of sharks and used by furniture-makers as a kind of sandpaper or put on the handles of swords to prevent the hands from slipping on the blood) and offered a reward if fishermen could bring him a specimen. The two fish caught in 1868 were 6.1m (20ft) and 5.5m (18ft) long, but Wright had also heard about fish that were supposed to be over 21m (70ft) long. The shark has been little studied and not much is known about it. Most of the information we have today is from a lifetime's research by American ichthyologist E W Grudger, who never saw the fish alive and only saw one specimen dead. Nevertheless, he collected many of the stories and pictures on which whale shark lore is based today.

Information on one piece of behaviour which Grudger was able to collect from Captain R W Mindte, who sailed in Californian waters, was just the kind of thing that might give rise to sea monster stories. Normally, a whale shark swims through the water, opens its large broad mouth, and sieves plankton and small fish from the surface waters. Occasionally, according to Mindte, a shark will float tail down in a vertical position with the mouth about a foot below the surface and by opening the mouth vertically suck in any small fish that are swimming by.

The whale shark lives in the warmer waters of the Atlantic, Pacific and Indian Oceans. The farthest north any have been seen is in the Gulf Stream off New York.

It is a harmless plankton-feeder and therefore is likely to be encountered close to the surface where it could easily be mistaken by an untrained observer for some unknown monster. To the trained eye, however, the whale shark is quite distinct. The brown-coloured skin – the thickest skin of all the sharks – is covered with white spots and the mouth is a wide slit at the front of a broad head. Many have since been caught or rammed by ships, with an average length of about 9.8m (32ft).

Real giants have been claimed; the passenger ship *Maurganui* hit a whale shark 97km (60mi) north-east of Tikehau Atoll in the South Pacific. The ship sliced into the shark's body with 4.6m (15ft) on one side of the bows and 12.2m (40ft) on the other, making 16.8m (55ft) overall. That was in 1934. The same year the Grace liner *Santa Lucia* was bound for Cartagena on Colombia's Caribbean coast when one night a shudder went through the ship. The captain slung lights over the bows to see what they had rammed and saw a 12.19m (40ft) long whale shark. Its bulk slowed down the ship noticeably.

Off the coast of Honduras, 'Sapodilla Tom', a 20m (66ft) monster, was seen regularly over a fifty-year period, and in Mexico's Bahia de Campeche, to the west of the Yucatan Peninsula, 'Big Ben' was reputed to be over 22.9m (75ft) long.

The largest whale shark scientifically measured, according to the *Guinness Book of Animal Facts and Feats,* was a specimen caught off Baba Island near Karachi, Pakistan. The body length, from the tip of the snout to the notch in the tail, was recorded as being 12.65m (41ft 6in), and its weight was estimated to be about 21.5 tonnes.

The skin of the whale shark is very thick and tough. It is covered in tiny teeth, known as dermal denticles, giving it the texture and strength of steel. The muscles below the skin can be tightened, making them as impenetrable as a truck tyre. Indeed, some observers have likened it to a steel-braced tyre. Harpoons, rifle bullets, or shot just bounce off. It makes the shark invulnerable to just about anything except an ocean liner. Indeed, the whale shark shows a remarkable tolerance to man.

For scuba divers, one of the ultimate underwater thrills must be to ride on the back of such a creature, and many films of divers doing just that have been seen on television in recent years. One characteristic the creature shows during these encounters is to dive down, every so often, into the depths. Why it should do this is not clear, but it can certainly go a long way down.

William Beebe, the underwater explorer, was in the Gulf of California in 1938 and came across what was estimated to be a 12.8m (42ft) whale shark. He tracked the fish for several hours. Knowing the skin to be tough, Beebe ordered two of his crew to jump onto the creature's back and ram home harpoons with a line and oil drum attached. The two men

> . . . made a beautiful pole-vaulting dive, with the harpoon between them. They struck hard and then leaped into the air and let their whole weight bear down, driving the harpoon home.

Understandably the inoffensive shark immediately crash-dived, returning to the surface some fifteen minutes later. The drum, still attached, had been crushed like an hour-glass indicating that the shark had been down to where the pressure is great. More harpoons just bounced off the shark's hide. Such an indifference to man's attacks prompted writer Frank Lane to speculate whether the whale shark rather than the whale was the subject of Job's inquiries in the Old Testament of the Bible:

> Canst thou draw out leviathan with a hook? or his tongue with a cord which thou lettest down? Canst thou put an hook into his nose? . . . Wilt thou play with him as with a bird or wilt thou bind him for thy maidens? . . . Canst thou fill his skin with barbed irons? or his head with fish spears? He esteemeth iron as straw, and brass as rotten wood. The arrow cannot make him flee: slingstones are turned with him into stubble.

Being unafraid of man and his boats and weapons the whale shark can, under certain circumstances, cause some consternation particularly among fishermen; it has the habit of scratching its back against the bottom of boats. Tunny fishermen in Californian waters are often visited and sharks can be seen with their backs coloured by the paint they have rubbed away from the bottom of boats.

With such bulk and tough defences, it is hard to imagine what might prey on such a creature, but in a letter to Gerald Wood (author of the *Guinness* book), Kendall McDonald told of the tail of a large whale shark having been washed up on a beach near the village of Dhahab on the Gulf of Aqaba, at the north-east end of the Red Sea (and site of a claimed 18.3m (60ft) whale shark in 1983). It looked as if it had been cut off by a knife, but McDonald noticed teeth marks on the portion that was left. Several creatures have the dentition to make a clean cut – sharks and killer whales, for example – but to slice into such a large creature would need quite a sizeable predator. It is unusual to find killer whales in the Red Sea and sharks made a clearly identifiable crescent-shaped wound. What could it have been? Is there some other gigantic creature lurking in those parts?

On the recent 'Reefwatch' programmes from BBC TV, viewers in the UK and USA were able to see 'live' the programme's underwater presenter feeding hard-boiled eggs to an enormous Napolean wrasse on a Red Sea coral reef. Perhaps an even larger relative nipped off the whale shark's tail. But, there again, it could have been something even more eerie.

Some weeks after the broadcast of *The Great Sea Monster Mystery,* I received a letter from George Whiting, skipper of the trimaran *Swingalong,* based in Cyprus. Man and boat had been in the Red Sea and in the tropical evening twilight of 21 February 1987 they came to anchor at 15° 33′N, 41° 50′E. They were just to the south of

Jabal Attair in the southern section of the Red Sea. The anchorage was about 9m (30ft) deep and was opposite a yellow bluff described in the Admiralty Pilot. Later that night something strange happened.

> It was a clear night, no moon, very black and no wind. Sea glassy. About 22.30, I went into the cockpit to have a look round before turning in. To the west, I saw a light patch which could have been a white hull, but as I watched, it moved toward me and I could now see that it was phosphorescent green. It seemed to be moving quickly; when close to the *Swingalong,* it appeared to be a disc about 40ft in diameter, flat on the surface of the water. The centre was a little less bright than the rest of the disc. The edges were quite sharp and seemed to be pulsing slightly. I stood watching if for about two or three minutes, and then it moved away and quickly disappeared in the dark. There did not seem to be any disturbance in the water as a result of its movements.

'What could it be?' George Whiting wanted to know. He suggested the shape was not one creature but many, say, a mass of mating squid. Might it though, have been the giant that chomped on the whale shark? We shall never know; truly a mystery.

Of all the sharks, the one most likely to be mistaken for a sea monster is the basking shark *Cetorhinus maximus,* the second largest living fish and the largest fish frequently encountered at sea or seen from the land. The longest individual to be measured accurately was a specimen trapped in a herring net in the Bay of Fundy, Canada, in 1851. It was 12.27m (40ft 3in) long and weighed 16 tonnes. That, however, was exceptional. The average used to be about 7.9m (26ft), but that was until man began to wipe out basking shark populations.

The largest of the basking sharks are the females and they come to the surface at times when food is abundant, when plankton blooms appear. From the scars on their bodies, it looks as if they have recently mated and are probably feeding up to supply the growing embryo with nutrients. While at the surface, they can be harpooned, hauled out, cut open, and their livers removed to extract the fine oils. With breeding females removed, a population quickly crashes, just as it did near Achill Island off the north-west coast of Ireland.

Like the whale shark, the basking shark is a harmless filter feeder. It has an enormous mouth which it holds open while skimming surface waters for plankton. Large gill-rakers trap the food particles while the filtered water leaves through large gill slits which almost encircle the body behind the head.

At the surface, the animal swims in leisurely fashion, often with the longish snout, dorsal fins and enormous tail protruding. It could easily be mistaken for something more unusual, particularly when it leaps clear of the water.

Occasionally, large numbers of basking sharks have been seen to swim closely together in line astern. Shark-catcher Fitzgerald O'Connor describes such an event in his book *Shark-O!*. The snout of each shark was raw and bleeding as it rubbed on the skin of the shark in front. Again, such a sighting by an unexperienced observer could give rise to more reports of sea monster sightings, such as those describing many dorsal fins in a line.

But the basking shark's main claim to monster fame is not achieved so much when the animal is alive but more likely when it is dead. Basking shark bodies, washed up on the shore, have hit the headlines on many occasions.

The most famous of the sea monster carcass stories took place at Stronsay in the Orkney Islands, and it illustrates how something quite ordinary can be misinterpreted and blown up out of all proportion. The source of information is the *Memoirs of the Wernarian Natural History Society*.

The story began on 26 September 1808 when John Peace and his crew, from Doonatown in Rothiesholm, were fishing near the eastern part of Rothiesholm Head. They noticed a flock of seabirds causing a commotion on what looked like the body of a dead whale and went closer to take a look. Peace's account was sworn in front of two Orkney magistrates. It recorded:

> That he then observed it to be different from a whale, particularly in having fins or arms, one of which he raised with his boat-hook above the surface of the water: that this was one of the arms next the head, which was larger and broader than the other nearer the tail; and at that time the fin or arm was edged all around, from the body to the extremity of the toes, with rows of bristles about ten inches long, some of which he pulled off and examined in the boat.

Watching Peace's boat from the shore was tenant-farmer George Sherar and he was able to confirm John Peace's account. About ten days later, a storm blew up and the strong easterly winds washed the carcass ashore. It lay on its belly just below the high tide mark. Sherar measured the body and found it to be 16.8m (55ft) long. The head was seen to be small and the neck about 4.6m (15ft) according to Sherar – but only 3.1m (10ft) according to a local carpenter. The bones were 'gristly' and the skin grey.

An artist came to draw the creature, but another storm broke up the carcass. Sherar, though, drew an outline of the body with chalk and the artist made his sketches based on the description given to him by local people. Several of the eyewitnesses confirmed that the very strange creature that he had drawn, which had six legs and a mane running the length of its body, was a true representation of the 'Stronsa beast', as it became known.

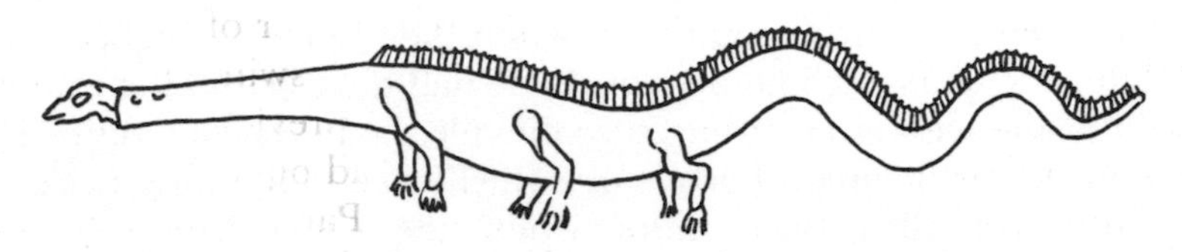

The Stronsa beast of 1808. *From a drawing by Mr Petrie.*

Fortunately, Sherar had kept some pieces of the animal, including the skull and several vertebrae. These were sent to the mainland and were readily identified as the cartilaginous remains of a shark skeleton, probably (although it was not confirmed conclusively) that of a basking shark. The identity debate, however, raged for many years. In the meantime, other carcasses were uncovered.

In the spring of 1885, according to an article in *Century Magazine*, the Reverend Mr Gordon of Milwaukee, President of the United States Humane Society, was on board a boat in New River Inlet on Florida's Atlantic coast. On raising the anchor, the crew discovered it had fouled on a long and heavy carcass that Mr Gordon first took for a whale but later described as saurian (dinosaur-like animal). It too had a long thin neck but the head was missing. At what appeared to be the anterior end of the body were two paddles. Mr Gordon thought he had found a recently living plesiosaur with a body length of about 12.8m (42ft). The creature was hauled out of the water to be taken away for analysis but, unfortunately for Mr Gordon, another storm meant that the sea took back the evidence.

The headless carcass found at New River Inlet, Florida in 1885.

Many years later, on 28 February 1934, another stranding was to reveal how 'prehistoric monsters' are really made. The place was the beach at Querqueville, to the west of Cherbourg on the Channel coast of France. Some fishermen found a decomposing carcass which had a one metre (3ft) long slender neck, a small head like that of a camel, two big flippers at the front of the 9m (30ft) long body, a fin on its back, and a tapering tail. The skin appeared to be covered with short whitish hair. Near the body were some of the internal organs, including what were described as the monster's 'lungs and kidneys'.

The Press had a field-day. Every conceivable expert was asked for his opinion and identifications ranged from a sea cow to a whale to a seal. Only one expert plumped for basking shark.

The event was further confused when the skipper of tugboat *117* told the newspapers about a creature he had seen swimming rapidly about in the sea not far from Querqueville the previous month. He described it as having a horse's or camel's head on a long neck.

Sanity prevailed once more when the Paris Museum was informed and Dr Georges Petit was sent to look over the body. He reached it on 3 March and found a grey, shapeless mass with skeletal column extending out in the front. He saw immediately that it could not be mammalian, and that it was indeed the remains of a shark. The anterior dorsal fin survived and the positions of the posterior dorsal, ventral and anal fins could be seen. Some intestines still lay on the sand near by and he could recognize the spiral valve characteristic of sharks and rays. He took specimens back to Paris and was able to confirm that the carcass was that of a shark, most likely a basking shark.

Examination of this and other subsequent basking shark carcasses revealed how the long-necked marine plesiosaur-like shape is achieved. The tissues around the very long gills often breaks apart first and the entire gill apparatus and lower jaw falls away, leaving the cranium shaped like a camel's head and the cartilaginous spinal column resembling a long and slender neck. Only the upper part of the tail fluke contains the end of the spinal column so the lower part breaks off, giving the creature the appearance of having a long thin tail. And finally, when the skin is sloughed away, the muscle fibres fan out like stiff white whiskers, giving the creature a 'furry coat'.

Imagine, then, the amusement afforded French scientists when their British counterparts tried to identify a 9m (30ft) long carcass that had been washed ashore at Girvan on the Firth of Clyde in 1953. It had a camel's head, a slender neck and a long tail which was covered in white bristles. The press understandably preferred a Loch Ness Monster interpretation, but the scientists, seemingly blissfully unaware of previous experiences, declared it to be a whale!

And indeed there had been many corpses washed up previously on Scottish shores. Hunda and Deepdale Holm were the recipients of monster flotsam during the Christmas period in 1941. The carcasses were formally identified by two scientists who preferred to remain anonymous as 'of the shark family – probably basking sharks'. But, despite the expert opinion, the local people were still sceptical. After all, they saw the long neck, the pointed tail, and the hair on the body resembling 'coconut fibre in texture' – all features that added up to a creature that looked nothing like any basking shark they had experienced before. Nevertheless, from the French experience, the bodies were showing the classic decomposition characteristics of a large shark.

A similar 'sea monster, 25ft long and weighing a ton', was washed ashore in Thurso Bay, not far from Dounreay on the north coast of mainland Scotland. No, it was not a mutant bi-product of the nuclear industry for it was reported in the *Standard* on 31 January 1945. 'No expert', the report said, 'has been able to identify the creature, which has a long tapering neck and a small head. The lines of the neck and the body are similar to those of a swan.' The dimensions and shape, however, sound suspiciously like the decomposing corpse of a basking shark.

Two rather strange carcasses washed up in Australia might also be identified as decomposing basking or similar shark carcasses. The first came to light with a letter to the Sydney *Sun* in 1948. The corpse, which was washed ashore on Dunk Island, Queensland, was described as being jellyfish-like, with no eyes and a tough skin with fur. It was apparently enormous, weighing several tonnes. The significant feature, however, was the presence of several large slits. Could these have been the remains of a basking shark's gill slits? The body caused quite a stink in more ways than one, and the local island folk resorted to dynamite to break it apart before moving it piecemeal out to sea.

The second unusual carcass arrived on the beach at Temma, on the west coast of Tasmania, in August 1960. After a heavy storm, a flat mass covered with what appeared to be short, stiff hair, and possessing gill-like slits along the side of the body, was found in the sand by local cattlemen. The tide buried it and exposed it many times during the next year or so, and visiting scientists came to no firm conclusion as to its identity. One did suggest, though, that it might be the body of a creature thought to live in the deep sea caves off Tasmania. On the other hand, could this too have been our old friend the basking shark?

Basking sharks, it is thought, account for over 90 per cent of sea serpent carcasses, but despite the knowledge accumulated recently, the sea serpent stories persist, even to this day. In 1970, for example, a 9.1m (30ft) long 'sea serpent' was washed ashore at Scituate, Massachusetts. It showed the usual camel's head and long thin body, and was identified later as a basking shark. But the carcass that caused the most excitement in recent times was hauled out of the sea by a Japanese fishing boat.

In April 1977, off the Pacific coast of New Zealand's South Island, the crew of the *Zuiyo Maru* hauled in their trawl net and discovered they had caught a monster. The 9.8m (32ft) long body was in an advanced state of decomposition, but they could see clearly the front and rear flippers on a creature that was 'like a snake with a turtle's body'. It was photographed and sketched and then thrown back into the sea lest it ruin the rest of the catch. Yokohama University and other scientists studied the photographs and rejected large fish and giant seals, plumping for plesiosaur as the most rational explanation. Fortunately, some tissues from the beast had been kept, and these were analysed by a biochemist at Tokyo University. They contained elastodin, a type of protein only found in the tissues of sharks and rays. Most likely, the fishermen had hauled up a decomposing basking shark which was rotting away in the usual 'sea serpent' way.

Other large sharks that appear from time to time and cause a stir include those from the deep sea. The Greenland, sleeper or gurry shark *Somniosus microcephalus* is the largest shark species after the whale, basking, and great white sharks. It can grow to a length of up to 7.6m (25ft) and was thought to be a rather lethargic creature. Food remains in the stomachs of hooked specimens, however, have ranged from fast-swimming creatures such as salmon and seals and even an entire reindeer (minus horns) so the sleeper can evidently put on a fair burst of speed when necessary. Whether it visits the surface waters and is a likely sea monster candidate is not clear. A sleeper shark, however, certainly caused a stir in Whitby when one Monday afternoon in the 1930s the fishing boat *Success* (which, on a previous occasion, had already brought in a world record 2.7m (8ft 9in) long tunny) arrived in port with a 4.6m (15ft) long specimen. When it was hauled up on to the quay the tightness of the wire hawser caused it to disgorge a large cod, a dogfish and a 1m (3ft) long porpoise.

An even bigger 6.4m (21ft) sleeper was caught off May Island in the Firth of Forth in January 1895. When the stomach was opened up, a seaman's boot complete with a leg was found inside. A 4.6m (15ft) specimen was trapped in nets in the Moray Firth in 1929. After quite a struggle to get it aboard the fishing boat *Bon Ami,* the shark disgorged four seals, one of which was over a metre long.

Closely related to the sharks are the skates and rays. One species of ray firmly in the giant league is the manta or devil ray *Manta birostris*. It can grow to about 7m (23ft) across the 'wings' and has been seen to leap clear of the water, perform somersaults, and re-enter the water with an enormous crash.

The manta ray is unusual among rays in that it swims near the surface in mainly warm seas. Most skates and rays are found on the bottom. Apart from its giant size, it is characterized by a lobe on either side of the head which serve to guide food into its large slit-like mouth.

It is quite possible that the tips of the fins protruding above the water, or the wide body itself, seen at the surface, could be mistaken for something more sinister. Similarly, other large rays, such as the bottom-dwelling cow-nosed, eagle or butterfly rays, all of which can grow to over two metres across, could be misinterpreted when seen by inexperienced eyes.

Most of the large sharks and rays are candidates for the more bulky sea monster sightings, but there is one shark which could be implicated in the great sea serpent story.

The frilled shark, *Chlamydoselachus anguineus,* is an ancient shark, almost a living fossil, which was found off Japan and identified for science by Samuel Garman in 1884. It has rarely been seen, but those that have been caught – from deep water off Japan, western Europe, California and south-west Africa – indicate a wide distribution.

The shark is peculiar in being eel-shaped. It has a single dorsal fin set far back on the body, a mouth at the front of the head (rather than underneath like most sharks) and six frilly gill-slits, the first of which almost encircles the body. Its appearance – the serpentine body and frilled 'mane' – has stimulated several commentators to suggest it as a candidate for the great sea serpent; that is, if *giant* frilled sharks exist in the world's oceans. The largest to date has been only 2m (6ft) long, but in ancient times the family had greater diversity with even larger species represented. Might some have survived, like the frilled shark itself, to the present day? There were two incidents that suggest this might be so.

The frilled shark-like fish seen by Captain Hanna in 1880.

In 1880, fisherman Captain S W Hanna of Pemaquid caught a strange eel-like fish in his nets. It was about 7.6m (25ft) long and only 25cm (10in) in diameter. The head was relatively flat, the nose stuck out only an inch or so above the mouth which contained sharp teeth. The skin did not have scales but was more like that of a shark. Although the fins were not quite right for a frilled shark there is no doubt that it was a largish relative of *Chlamydoselachus*.

Bernard Heuvelmans found another reference in the *Shipping Gazette* for 1886. It read:

> A sea serpent is reported to have been captured at Carabelle, Florida, by a fishing steamer named the *Crescent City,* which it towed wildly for some time before it was killed. The thing measured 49 feet long and six feet in circumference. It is eel-shaped, with a shark-like head and a tail armed with formidable fins. It was caught with a shark-hook, but after being tired out it had to be shot.

Here, at last, is one candidate for the identity of the great sea serpent. Another favourite for the title is a bony fish, in fact, the longest bony fish in the world. It is called the oarfish or 'King of the Herrings', *Regalecus glesne,* and it has several features that fit with eyewitness descriptions.

The fish can be very long and eel-like. Scientists from the Sandy Hook Marine Laboratory, taking part in a research cruise on the research vessel *Challenger* on the 18 July 1963, claim that they saw a 15.2m (50ft) oarfish in the sea near Asbury Park, New Jersey. Another giant had its head sliced off by the bow of the *Santa Clara* as it steamed past the North Carolina coast on its way to New York. The fish was thought to be about 13.7m (45ft) long.

Those that have been measured include a 7.6m (25ft) specimen taken off Pemaquid Point, Maine, in 1885 and a 6.4m (21ft) fish washed ashore at Newport Beach, California, in 1901. Several smaller specimens have been found elsewhere in the world – off the north-east coast of England and the Pacific coast of New Zealand – indicating a worldwide distribution. The fish is not commonly seen though, and probably lives in mid-waters, rarely coming to the surface. If, however, it did poke its head above the water, any observer would likely have quite a fright.

On the oarfish head is a crest of long spines, usually bright red in colour, which can be erected into a crown-like plume. The paired pelvic fins consist of long spines that end in broad, feather-like 'oars' – hence one of its names. Down the entire length of its silvery, ribbon-like body (flattened dorso-ventrally) is a long, spiky dorsal fin. It is certainly a startling-looking fish, and given its size and shape – a seven-metre specimen is likely to have a metre-high crest

– it could easily fit the description of a 'horned' sea serpent. One would think, though, that if presented with an oarfish, observers would draw attention to the bright colours (unless seen in silhouette); and this has not been the case. It also does not have the musculature capable of supporting the head out of water. Indeed, when caught, the body has been found to break up easily. Nevertheless, I feel that it should not be dismissed as a candidate. An Australian sea fisherman at Waitpinga Beach, near Adelaide, certainly did not dismiss his encounter with the fish. The story was sent to me by Professor C C von der Borch of the Flinders University of South Australia:

> A salmon fisherman, who was surf fishing, hooked a large (several metres long) oarfish. This is normally a deep sea species, but must have entered the shallow shoreline zone . . . As the luckless fisherman reeled in his catch, expecting to see a salmon on his hook, a large wave caught the oarfish and it 'surfed' on to the beach. The fisherman suddenly saw the apparition of a sea serpent, apparently coming at him out of the depths. He dropped his rod and fled to the sand dunes. Remains of the oarfish subsequently found their way to the School of Biological Sciences at Flinders University.

Whitby decorator Robert Wood was equally startled by an oarfish that was washed up on Whitby beach a few years ago. He was taking a dusk stroll when his dog started to bark at something lying on the beach. It turned out to be a 3.8m (12ft 6in) oarfish. He took it home in his decorating van where it was recognized by a Ministry of Agriculture, Fisheries and Food representative. Mr Wood later donated the fish to the British Museum (Natural History).

And, in the not too distant past, the scientific community was taken in by a report from Bermuda. It was in January 1860 that the learned journal the *Zoologist* published a story entitled, 'A Sea Serpent in the Bermudas' by Captain Hawtaigne RN. It was, so the article claimed, the same creature that featured in the *Daedalus* affair in 1848 (see page 107).

On this later occasion two men were gathering seaweed beside Hungary Bay on Hamilton Island, Bermuda, when they heard the sound of disturbed water. They ran to the spot and discovered a large creature stranded on the rocks. It was flailing about but was unable to extricate itself and return to the sea. The men took their seaweed forks and killed it.

> The reptile was sixteen feet seven inches in length, tapering from

> head to tail like a snake . . . the colour was bright and silvery . . . But its most remarkable feature was a series of eight long, thin spines of bright red colour springing from the top of the head . . .

At the time, the report caused a great deal of excitement and it was claimed that the men in Bermuda had captured a sea serpent. Some time later, a *bona fide* naturalist was able to examine the carcass which had, incidentally, been left to rot on the shore where it had been killed. Without hesitation he recognized that the 'sea serpent' was none other than the oarfish. There were, without doubt, a lot of red faces all round.

The oarfish washed ashore at Hungary Bay, Bermuda, in 1860. *From an engraving.*

A relative of the oarfish is the dealfish *Trachypterus arcticus*. It has a long eel-like body with a large orange dorsal fin running down the back. It is slightly smaller than the oarfish with maximum known lengths of up to 3m (10ft). It has rarely been seen alive, but is sometimes washed up on North Sea coasts.

The largest bony fish in the sea, and again one which might figure in sea monster sightings, is the ocean sun fish *Mola mola*. Individuals, which look like huge vertically swimming dinner plates, have been found in most oceans, and they can reach immense sizes.

The biggest sun fish so far recorded was encountered off Bird Island about 64km (40mi) from Sydney, Australia, in September 1908. The steamer SS *Fiona* was heading for port when the entire ship shuddered as if it had hit wreckage. The port engine packed in entirely. The captain, thinking the port propeller was snagged, ordered the starboard engine to be shut off too. A boat was lowered to check for damage.

As the inspection party reached the port propeller, they could see that an enormous sun fish had lodged in the propeller shaft support and a propeller blade was firmly embedded in its body. It was impossible to remove the creature at sea and so the ship limped home on one engine.

At Port Jackson, the ship was anchored in Mosman Bay and the fish removed. It was taken ashore and weighed and measured. It was 3.1m (10ft) long, 4.3m (14ft) from the top of its dorsal fin to the bottom of its anal fin, and weighed a staggering 2,235kg (4,927lb).

Most of the ocean sun fishes found in recent years are much smaller than the Bird Island monster. They are not commonly seen, but are sometimes found floating on their sides at the surface. Ichthyologists have speculated whether this is 'basking' behaviour, but now it is generally thought that these are specimens about to die. Young fish have been seen to swim in an upright position, sculling rapidly with the two large fins.

I was fortunate enough to have examined a 2m (6ft) sun fish washed ashore near Taormina, Sicily, and the first thing I noticed, apart from the disc-shape of the blue-grey body, was the tiny mouth with which it eats small jelly-fish and other soft-bodied drifting life.

Like the whale shark, the sun fish has a thick layer of gristle under its smooth skin. This affords it ample protection against harpoons and even rifle bullets, and in its natural state protects against attacks from sharks and the like. These two giants, living a similar lifestyle, have adopted the same strategy for survival.

As for its status in the sea monster stakes, any large and unusually shaped creature, lolling about at the surface, is likely to be called into question and even the most expert of observers, with only a brief glimpse of an animal, can be puzzled.

One of the biologists from the Marine Biological Association's laboratories at Plymouth was on the research vessel *Discovery* in the early 1950s. The ship was in the vicinity of the Azores during bad weather; the crew were below eating lunch, and the biologist was stowing gear away on deck in the stern. He chanced to look over the side and saw a greyish animal, about 3.1m (10ft) long and 2.4m (8ft) across, just below the surface of the water. He was able to watch it for a few minutes through binoculars, and could eliminate a giant ray, a whale, a sea turtle, seaweed or wreckage as plausible explanations. He rejected sun fish on the ground that he thought his 'private monster' too large, but those who have heard the tale are sure that this is just what he saw.

Experts of a different sort – two Scottish boatmen and ex-shark fishermen – encountered a strange fish off the Scottish west coast in August 1950. Bruce Watt and John McInerney, who caught basking sharks with Gavin Maxwell, were taking out trippers in their boat the *Islander*. In Maxwell's book *Harpoon at a Venture*, he describes how Watt and McInerney were in the mouth of Loch Hourn, on the south side near Crolag, when they spotted what they thought was a basking shark's dorsal fin. They approached slowly, hoping to show the holidaymakers the giant fish. But, as they got closer, they noticed the fin was unlike that of the basking shark – it was paper thin; closer still they could see the rest of the body underneath the water. Watt described what he saw:

> I expected to see fifteen feet or more behind the fin. Instead of that there was only five or six feet behind it, and in front of the fin it seemed to end off short in a sort of egg-shaped nose, in which I thought I could see a large eye. The tail must have been very small . . . and it certainly was not a powerful-swimming fish like a shark. I'd say its depth was at least half its total length – a very squat kind of fish. It looked very dark, darker than a shark, and the whole fish wasn't more than about eight feet or so.

The creature Watt had described, and which was 'quite outside his long and wide experience at sea', was undoubtedly a large specimen of the ocean sun fish – an unusual visitor to British waters but,nevertheless, a known one.

The ease with which very ordinary sea creatures can acquire the proportions of sea monsters is ably demonstrated in the encounter recorded by Norwegian biologist Alf Danniveg. He wrote to Maurice Brown of the BBC in November 1959:

> I was out fishing in the skerries a fine summer morning. The sun was just rising above the horizon. No wind stirred the surface, only a small swell from the foregoing day drifted along the skerries.
>
> Then suddenly, I was aware of 'something' breaking the surface half a mile further out towards the ocean. My first impression was that a small whale had made some ravages into a herring swarm. But then it turned out that the breaking and splashes moved towards land. And I was able to state an undulating vertical movement.
>
> I was convinced that it was a large creature swimming at a great speed near the surface. I was not able to identify the animal – and the possibility of its being an unknown species struck my mind. Finally, I was quite sure. It was an eel-like large animal, a sea serpent!
>
> The animal came nearer and passed me at a distance of perhaps

> 50 fathoms. Then I perceived in the foam the contours of a large salmon – pursued by a seal.
>
> Thus ended the story of the sea serpent.

The salmon, even the largest (which may grow up to 1.5m (5ft) in length), are small fry compared to some of the giants in the bony fish world. Giant groupers, for example, have been known to swallow whole an oil-rig diver – lead boots, helmet and all. And if that seems an exaggeration, diver J T L Ponds can assure us that it is not. He and a colleague (Jake) were diving on an oil-rig in the Gulf of Mexico and were being harassed by a particularly large grouper; it was 'big enough to swallow a Volkswagen', according to Ponds. Every time it opened its mouth a strong current would draw in fish and as it approached the divers they could feel the movement of the water.

> Suddenly I seen Jake's feet sticking out this here fish's mouth . . . So I grabbed his lifelines and started to pull hard. Nothing happened. That fish just remained there without so much as blinking one of those big eyes of his. That gave me a sudden inspiration. I hit him square in the eye with one of my fists. He just spit Jake out.

Groupers can be over 2.4m (8ft) long and weigh nearly half a tonne.

Another bony fish with a hearty appetite for human flesh is the giant barracuda, a thin, pike-like fish that can grow up to 3.1m (10ft) long. Barracudas have been known to enter shallow water and sever the leg of a victim paddling in the shallows. Some people consider it to be the 'most formidable bony fish in the sea'. Certainly the power with which the fish can strike is illustrated by an incident that took place in the Windward Islands.

A seaman taking part in an offshore rowing race had painted the ends of his oars with gleaming white paint. He was a strong rower and was approaching the winning line when one of his oars was ripped from his grasp and disappeared below the surface. Fortunately for the oarsman, he could recover the oar, and firmly embedded in the blade were the fang-like teeth attached to a 1.2m (4ft) long barracuda. Apparently, it made delicious eating.

Fast-swimming swordfish, sailfish, marlins and tuna can grow to immense sizes. British records give a 6.1m (20ft) maximum length for the swordfish. They are also some of the fastest fish in the sea, an attribute that can be a mixed blessing, for they have, it seems, great difficulty in stopping. There are several instances of ships getting in the way, including the research submersible *Alvin,* which was holed by swordfish. At London's British Museum (Natural History) there is a piece of a ship's hull 56cm (22in) thick, that is pierced by a 13cm (5in) diameter sword. The fish itself must have been enormous. Given that swordfish and their relatives have tall fin crests behind the head and large dorsal fins extending along the back, they too must be included in the list of creatures that could be misidentified at the surface.

Of the flatfishes, the halibut *Hippoglossus hippoglossus* can grow to a huge size. The largest specimens have been over 4m (13ft) in length and weigh over 300kg (661lb), although nowadays, it is rare to find halibut over 2m (7ft) long. They live at depths down to 2,000m (6,560ft) along the edge of the continental slope from the Faroes to Greenland and around Newfoundland, but occasionally are found free-swimming and in shallow water. There is a closely related species living in the north Pacific.

One fish that we should not ignore in considering candidates for sea monster reports is the sturgeon. Some forms of this ancient fish, such as those found in the northern Adriatic, the Black Sea, Volga River, and Danube can grow to 9m (20ft) long, but the more normal length for an Atlantic sturgeon, *Acipenser sturio,* is 1 – 2m (3 – 6ft), with a maximum of about 5m (16ft). It is a bottom-feeding fish with a shovel-nose equipped with sensory barbels. It spends much of its time at sea but enters rivers to spawn. In the 'sea monster' context its most interesting feature is the armour-plating. Several sea serpent reports include reference to rows of triangular projections sticking up out of the water. The sturgeon has just such a row of large armour-like scales that run the length of the back.

One such armour-plated sturgeon caused considerable consternation in the River Tywi at Nantgaredig on the A40 near Carmarthen, south-west Wales, on 28 July 1933. The creature caused 'enormous waves' in the river and a local fisherman, Alec Allen, managed to foul-hook it. He thought at first that he had hooked a log, but when the log began to thrash about and to leap out of the water he realized he had caught an enormous fish. He struggled with it for some time and eventually hauled it into shallow water. There he killed it by bashing its head with a large rock. A spectator on the bank, seeing the size of the fish and the struggle that Alec Allen was having, ran hysterically from the scene. The fish was 2.8m (9ft 2in) long, 1.5m (59in) wide, and weighed 176.2kg (388lb).

An even larger specimen was caught in the Beauly Firth at Kirkhill,

The Associated Press Ltd

Only three specimens of megamouth shark have ever been found. This specimen was washed up alive on a beach near Perth in Western Australia.

The Associated Press Ltd

The body of the largest known leatherback turtle in the world was washed ashore near Harlech on the Welsh coast. It can be seen on display at the National Museum of Wales in Cardiff near Inverness on the east coast of Scotland, in 1661. The sturgeon was over 3.7m (12ft) long. The largest known sturgeon in British waters was a 230kg (507lb) fish taken from the Severn in 1937. Others have been caught in the Conway River, below Tal-y-Cafn bridge, in the Wye at Chepstow, and, according to Travis Jenkins' *The Fish of the British Isles* published in 1925, 3.04m (10ft) specimens (and even a giant 5.49m (18ft) sturgeon) have been caught off Morecambe, north-west England, and Pwllheli on the Lleyn Peninsula, north-west Wales. Specimens have also been seen in the Severn at Shrewsbury, the Trent at Nottingham, and in the Thames.

Reports of sturgeons in British waters tend to be accurate for the fish is a 'Fishe Royale' – it belongs to the Crown. If caught it must be offered first to the reigning monarch. This custom was started in the fourteenth century by Edward II. Exceptions to the royal ruling were the Lord Mayors of London who were first in line for any sturgeons taken from the Thames.

The sturgeon is a valuable fish – the flesh is said to taste 'like a compound of veal and eel, with a flavouring of lobster' (a description also applied to the flesh of the wels or giant European catfish – a freshwater fish); substances from the swim-bladder are used to make isinglass; the dorsal region is smoked and dried as 'balyk'; and the roe is sold worldwide as caviare.

Reptiles at sea provide us with a whole host of sea monster candidates – giant anacondas, pythons and monitor lizards, living dragons and sea turtles, and salt-water crocodiles that are refugees from a bygone age – it is as if we have opened the pages of prehistory and the all the ancient characters have come alive.

The most awe-inspiring of the reptiles is undoubtedly the salt-water crocodile *Crocodylus porosus,* or 'salty', as it is known in northern Australia. Many animals have been found reaching lengths of 5.5m (18ft) and weighing 816kg (1,800lb). Salt water crocs live in the Indian Ocean, around the coasts of south-east Asia and northern parts of Australia. They have been found at sea far from land and are clearly creatures that could be taken for sea monsters as they really are monsters themselves.

Some of the largest crocodiles known have been found in the Segama River in north Borneo. Rubber plantation owner James Montgomery, concerned about the safety of his workers who bathed and did their laundry in the river, shot twenty crocs between 6 and 8m (20 and 26ft) long. Not far from his house one giant was left well alone. The local river people, the Seluke, considered it the 'Father of the Devil' and would throw silver coins into the river to appease it whenever it appeared. Montgomery was able to measure the imprint it had made while resting on a sandbank and he found it to be over 10.1m (33ft) long.

Another haven for giants appears to be Papua New Guinea. A 6.32m (20ft 9in) specimen was shot at Liaga on PNG's south-east coast and another 6.2m (20ft 4in) croc was trapped in a fisherman's net at Obo in the Fly River.

Perhaps the most famous 'salty' stories come from the Northern Territory of Australia, in particular, the tale of 'Sweetheart' in the Finniss River.

Until comparatively recently, Australians shot and skinned their crocodiles. The animals were understandably wary of man and made themselves scarce whenever man appeared on the scene. Today, salt-water crocodiles are considered a threatened species and killing them has been made illegal. The crocs consequently are less wary and some are positively bold. Sweetheart was one such giant that was not fearful of man or his boats. In fact, he (not she as the name suggests) found boats quite irresistible and took a fancy to outboard motors – much to the consternation of local sports fishermen.

Several boats were attacked by this 5.5m (18ft) giant which weighed 816kg (1,800lb), and the crunch came when it took the transom off a small boat, spilled the fishermen into the water, and caused one startled angler to do 'the best performance of walking on water in two thousand years'. Curiously, nobody ever got hurt, but the wildlife rangers were going to take no chances and decided to bring Sweetheart in. He was to be sent to spend the rest of his days in a crocodile farm. But, during the capture he drowned, and his body can now be found in the Darwin Museum.

The fear that a salt-water crocodile the size of Sweetheart would eat a human was well founded. Many people have been killed by these ferocious and unpredictable creatures. Recently, Australia was seized with 'crocodile fever' and every croc was threatened with a bullet between the eyes after a spate of human killings along the Queensland coast in the 1980s.

Two killings were just a couple of months apart. During Christmas 1985, a 43-year-old woman was paddling in a creek after a Christmas dinner party when she was eaten by a 5m (16½ft) croc. Only her fingernails, toenails and bones were found in the animal's stomach. And in February 1986, a 26-year-old woman was attacked and killed at Cape York. She and her companion had been stranded on a sandbank when their dinghy's outboard motor failed. They decided to swim for their larger boat. The man was just getting aboard and turned to help the girl when a giant croc seized her and carried her off. Her dismembered body was found later beside a

5.2m (17ft) long salt-water crocodile.

The daddy and the most feared of all living crocodiles, however, is Bujang Senang – the King of Crocodiles. He lives in the Lumpar River in Sarawak and is distinguished by a white patch on his back. He is thought to be about 7.6m (25ft) long, although few people have actually seen him; and those who have encountered the giant have not lived to tell the tale.

The local Ibad tribe hold the creature in great esteem. Crocodiles are deified and it is forbidden to kill them, except in self-defence. The deputy police commissioner for Sarawak is not a believer. He has been more concerned with the many deaths that have occurred along the Lumpar River in recent years. It is thought that all the attacks are down to one crocodile – Bujang Senang.

One of his latest victims was a fisherman who had been standing on the river bank talking with a friend. The crocodile left the water, thwacked the fisherman with his powerful tail, and dragged him screaming into the water. Now, the hunt is on, but Bujang Senang is proving very elusive.

There is no doubt that an encounter at sea with one of these giant 'salties' could give someone quite a shock. They have been seen near remote island groups, such as Fiji and the Cocos Keeling Islands, far from the mainland where they breed. Their prehistoric appearance has all the qualities needed by a sea monster.

Another giant reptile is the leatherback turtle. Many commentators believe it is this creature that has given rise to many of the short-necked, big-bodied sea monster sightings, such as those of 'Morgawr' off Falmouth. The turtle is global in its distribution, venturing from tropical to temperate waters, a giant vagrant following the ocean currents. The largest Atlantic leatherback was found washed up on the shore near Worthing in 1987. It measured just 13mm (½in) shorter than the world record-holder – a Pacific leatherback captured alive in Monterey Bay in 1961 which was 2.54m (8ft 4in) long and weighed 865kg (1,908lb). Another giant over 2m (6ft) long and weighing over 318kg (700lb) was landed by a lobster fisherman near Port Isaac in Cornwall on 10 August 1988. Several large specimens had been seen off the South Devon and Cornwall coasts. It is likely they were following the unusually large shoals of giant jellyfish – 'the size of dustbin lids'.

An even larger specimen was reputed to have fouled the fishing nets of the *Castle Moil* in the South Minch near the island of Barra (which has the only aircraft runway that is covered by the tide twice a day) in the Outer Hebrides of Western Scotland. When the fishermen attempted to winch the giant turtle aboard it broke the nylon rope and fell back into the sea. There was no accurate measurement taken but the crew estimated it to be 3.7m (12ft) long.

The leatherback turtle, unlike all other sea turtle species, is characterized by a smooth back with seven longitudinal ridges but without scales. Many other turtle-like monsters are said to have large scales, which rules out the leatherback as a candidate. No other known turtle reaches the giant size of the leatherback, so what could these other creatures have been?

During the Cretaceous period, when the dinosaurs ruled the earth, a huge sea turtle with a 3.7m (12ft) diameter carapace and a whole body length getting on for 7.6m (25ft) swam the oceans. It is only known from the fossil record and has the scientific name *Archelon*. There is some evidence that sea turtles of this size survived until quite recently. In the third century AD, Claudius Aelianus wrote in his *De Natura animalium* about 3.7m (12ft) turtle shells from the Indian Ocean that were used as roofs for huts. And, Al Edrisi. in his *Description of the World* published in AD 1145 describes 9m (30ft) long turtles found to the west of Sri Lanka. Could some of those creature still be alive today?

One of the problems associated with giant sea turtles is that they must come to land to lay their eggs; to date, as far as I am aware, nobody has described seeing anything of such size leaving the water, hauling out to the top of a beach, digging a hole, and laying its eggs.

Of the lizards, the Komodo dragon *Varanus komodo* is the largest (although not the longest – that honour goes to Salvadori's monitor *V. salvadori* from Papua New Guinea) with some specimens reaching over 3m (10ft) in length and weighing 160kg (353lb). It becomes a potential sea monster on account of its habit of swimming between islands. In the water, the Komodo dragon, unlike the salt-water crocodile, raises its head above the surface of the sea. It is quite possible that crocodile-like sea monsters could have been, in reality, monitor lizards.

The largest population of Komodo dragons is centred on the tiny Indonesian island of Komodo, but the lizard is also found on the neighbouring islands of Flores, Rintja and Padar. The largest specimen claimed was spotted in 1938 by an American journalist on the beach on Komodo. It was reputed to be about 4.42m (14ft 6in) long. The previous year a Swedish zoologist claimed he had seen a 7m (23ft) dragon. And, on the island of Flores two Dutch pearl collectors said they had killed several 7m (23ft) specimens. None of these measurements has been authenticated. The largest Komodo dragon with believable accurate statistics, according to *The Guinness Book of Animal Facts and Feats,* was a male specimen kept in the St

Louis Zoological Park around 1937. It was 3.10m (10ft 2in) long and weighed 165kg (365lb).

Komodo dragons can be dangerous. In the past few years two tourists are known to have been killed and partly devoured by dragons, and these reptiles have even been known to dig up fresh graves and eat the contents. Their normal food is carrion, although there are tales of Komodo dragons following pregnant goats and waiting for them to drop their babies.

Giant snakes are a possible explanation for the giant sea serpent, although the largest of the snakes, such as boa constrictors, reticulated pythons and anacondas, are rarely seen in salt water. They can, however, grow to sea serpent sizes.

Anacondas, *Eunectes murinus*, are the most likely candidates as very large specimens spend all their time with their enormous bulk supported by the water in rivers and lakes. The invading Spaniards told of 18 – 24m (60 – 80ft) anacondas which they called 'matatora', meaning killer of bulls.

There are many giant anaconda stories, but most must be taken with a good deal of scepticism. In 1948, for example, a 40m (130ft) anaconda was supposed to have been captured alive by local Indians beside the River Amazon. It was towed to Manaos where some lunatic, in true frontier style, shot it. Later that year a 35m (115ft) snake was killed at Abunda and in 1954 a 36m (118ft) specimen allegedly was shot at Amapa. And, in 1907, Colonel P H Fawcett shot an anaconda, which he claimed was 18.9m (62ft) long, on the Rio Abunda.

The largest accurately measured anaconda was one shot on the Colombia-Venezuela border in the 1940s. Members of an oil exploration team said that they spotted the snake in the Upper Orinoco River. They shot at the creature, dragged it from the water, and measured it. It was 11.43m (37ft 6in) long. They left the snake on the river bank while they continued with their work, but when they returned to remove the skin they discovered the snake was gone. It was not dead at all; it had been only stunned by the bullets.

Of the other large constricting snakes, a size of 39.6m (130ft) was claimed for an African rock python, *Python sebae*, killed by Bwamba tribesmen in the Semliki Valley, Central Africa, in 1932; an 11m (36ft) reticulated python, *Python reticulus,* is said to have been captured in Celebes as recently as 1971; and a 7.6m (25ft) amethystine python *Liasis amethistinus* was killed near Cairns, Queensland, Australia.

The most bizarre and frightening story comes from Sumatra where a bulldozer of a forest logging project disturbed two giant snakes. According to the newspaper reports, the driver of the machine had a life and death struggle with the two writhing monsters before he managed to crush one. It was claimed to be 25m (82ft 6in) long and inside were the bodies of four humans, two with their shorts still intact.

Clearly, even believably large snakes can be quite formidable and dangerous. Anacondas and rock pythons have been found with 1.5m (5ft) long alligators and crocodiles in their gut. Others have been found containing whole pigs. The largest intact item of food found in a snake was an impala inside a 4.87 (16ft) rock python. Despite the Sumatran tale, it is unlikely that a large snake could take a fully grown human. The body would have to be eaten head first and the shoulders would probably jam in the snake's mouth. There have been, however, several cases of babies or children caught and eaten by constricting snakes. In November 1979, for example, a python in the Transvaal killed a 13-year-old South African shepherd boy.

Pythons *have* been seen at sea, even though they do not normally frequent the marine environment. In the Indo-Pacific region, pythons, like the Komodo dragons, swim from island to island in search of food. It is reported that one of the first creatures to reach the remains of Krakatoa after the eruption in 1883 was a large reticulated python. It must have swum the 50 kilometres from Java.

Also, large constricting snakes could be accidentally washed out to sea, say, after a river had become swollen by heavy rains. There was a report in the *Zoological Journal,* dated December 1827, of a large snake having drifted on a tree-trunk to St Vincent in the West Indies from a South American river. The nearest major river, the Orinoco, is nearly 300km (186mi) away.

It has been speculated that Chessie (see page 64), could be a large aquatic constricting snake, such as an anaconda, but the reports of sea serpents have one observation that does not tally with the snake hypothesis: sea serpents seem to move consistently with a vertical undulation of the body, whereas snakes wriggle through the water with horizontal movements. The shape and structure of a snake's back-bone, with the row of neural spines on the top of each vertebra, also precludes excessive vertical movements. One explanation for the vertical movement is that the creature could be swimming on its side. All snakes can swim well using the horizontal undulations.

Sea snakes, that *do* live in the sea, have flattened, oar-shaped tails to aid swimming. It is unlikely that they could be mistaken for giant sea serpents because they do not grow very big. The largest known sea snake is the yellow sea snake *Hydrophis spiralis* which averages at 1.8m (6ft). Some specimens in excess of 2.7m (9ft) have been seen. Stoke's sea snake, *Astrotia stokesii,* which is a much bulkier sea snake, can grow to 1.8m (6ft). The sea snake most likely to be encountered at sea, and the only one which is entirely marine-living, is the yellow-bellied or pelagic sea snake *Pelamis platurus* (meaning flat-tailed wanderer), but it rarely exceeds a metre. Also, it lives only in the tropics, in the Indo-Pacific region, and cannot survive in sea temperatures below 18°C. Even in seas at a temperature of 17°C it cannot live for very long.

Often the yellow-bellied sea snakes are found in large aggregations in calm waters. They lie in the water and mimic sticks. Small fishes seek protection below the 'stick' and the snake simply swims gently backwards and grabs a victim, injecting it with venom and swallowing it whole. The yellow-bellies appear to have no predators. In an experiment, a large fish which would not normally encounter the snakes was fed some sea snakes. They bit the fish in its stomach, injected their venom and when the fish died they swam out of its mouth.

There is one sea snake observation, however, which *is* relevant to the sea monster story. The event happened in the early 1930s and is described in Willoughby Lowe's book *The Trail That is Always New,* published in 1932. He was on board a steamer which had reached the Malacca Strait between Sumatra and the Malay Peninsula. He was on deck with other passengers after lunch, and, looking towards the land, he saw a long line in the sea running parallel to the ship some eight kilometres away. He and the others were curious but thought no more about it until later. Having finished afternoon tea, they all came back on deck.

> On returning to the deck we still saw the curious line along which we had been steaming for four hours, but now it lay across our course . . . As we drew nearer we were amazed to find that it was composed of a solid mass of sea snakes, twisted thickly together. They were orange-red and black . . . Along this line there must have been millions; when I say millions I consider it no exaggeration, for the line was quite ten feet wide and we followed its course for some sixty miles. I can only presume it was either a migration or the breeding season . . . it certainly was a wonderful sight.

Lowe's snakes were probably Stoke's sea snakes, and they present us with one more natural phenomenon which could be mistaken, albeit on a smaller scale, for a mysterious sea monster.

Reptiles, with their rather prehistoric appearance and movements, are favourites for the identity of the great sea serpent, but when we come to the mammals we find an enormous assemblage of animals that could be mistaken for something else.

If we start with the gentle giants of the sea – the great whales – we find that these animals, though enormous, are relatively easy to identify. Whales have the large tail flukes which sometimes lift clear of the water before a deep dive or splash down in a mountain of spray when 'tail-lobbing'. There is the blowhole and characteristic 'blow' pattern for each species of whale. And there is the way the whale's back arches slightly as it swims at the surface. Nobody, it seems, should mistake a whale. There are, however, certain circumstances in which whales do not look like whales.

Dr Martin Angel, of the Institute of Oceanographic Sciences at Wormley in Surrey, told me about the time he was on board the research ship *Discovery* and spotted his first 'sea monsters'. They were minke whales with just their heads sticking out of the water – behaviour known as 'spy-hopping'. It is the way many species of whales and dolphins look at what is going on above the surface. Dr Angel said that the pale countershading of the underside of the head and throat made it look as if the top of the head was on a narrower 'neck'.

At the University of Manitoba, Dr Waldemar Lehn of the Department of Electrical Engineering and his colleague Irmgard Schroeder took the 'spy-hopping' explanation for strange marine phenomena a little further by studying optical distortions in the atmosphere under certain weather condition.

The researchers started their investigations after having read a mid-thirteenth-century manuscript, the *King's Mirror,* that describes the 'merman'.

> This monster is tall and of great size and rises straight out of the water . . . It has shoulders like a man's but no hands. Its body apparently grows narrower from the shoulders down, so that the lower down it has been observed, the more slender it has seemed to be. But no one has ever seen how the lower end is shaped . . . No one has ever observed it closely enough to determine whether its body has scales like a fish or skin like man. Whenever the monster has shown itself, men have always been sure a storm would follow.

Lehn and Schroeder were intrigued by the description and, digging a little deeper, they found another even earlier reference in the *Historia Norvegiae* written in 1170.

> . . . there is the whale, and there is the merman, the largest wild beast, but without head and tail, thus it is just like a tree trunk when it leaps up and down, and it does not show itself without its foretelling danger for seafarers.

Those were apparently the only traceable medieval references they could find. They were struck, though, by the fact that both referred to events taking place before storms. They decided to put the available information about atmospheric conditions prior to storms on computer and run a clever programme that duplicated stormy-weather images. They used as their subject matter a 'spy-hopping' killer whale and a walrus sitting upright in the water. The results were very interesting. Under certain conditions, both animals viewed from over a kilometre away appeared much taller than they really were, with a narrow 'neck' or 'waist' at the base. The killer whale was transformed into a credible sea monster and the walrus into a 'merman'.

The cause of the distortion is a temperature inversion that occurs when a mass of warm air moves over cold air. When light passes through the disturbed air, it is distorted in the vertical plane, i.e., it is stretched, but unaffected in the horizontal plane.

In the experimental programme Lehn found that a temperature difference of about 7.5°C with the boundary or thermocline about 2.2m above the sea surface was sufficient to distort the image of the killer whale. In those conditions someone with their eyes at a height of 2.1m above the sea, about the right height for a sailor standing on a Norse long-ship, would not recognize the whale as a whale but would more likely see a curious monster.

Interestingly, the weather conditions in which these effects occur naturally are at the last stages of the passing of a warm front, usually a calm period prior to a storm. The interface between the warm and cold air masses would then be right down close to the sea surface. If it is calm, the boundary would be fairly straight and images would be clear but distorted. The Norse seafarers, then, had not been making up stories. They genuinely thought they had seen strange creatures, and because of Lehn's work we can now believe them.

Lehn was also able to compare his computer study with the real phenomenon in nature. He was walking along the shore at Lake Winnipeg on a warm day, but at a time when there was still ice on the lake. Air temperature was about 28°C while water temperature was nearer 0°C. He spotted a 'merman' and took its picture. Walking just over a kilometre further on, he discovered his apparition was simply a small boulder.

The Canadian researchers were not the first to draw attention to

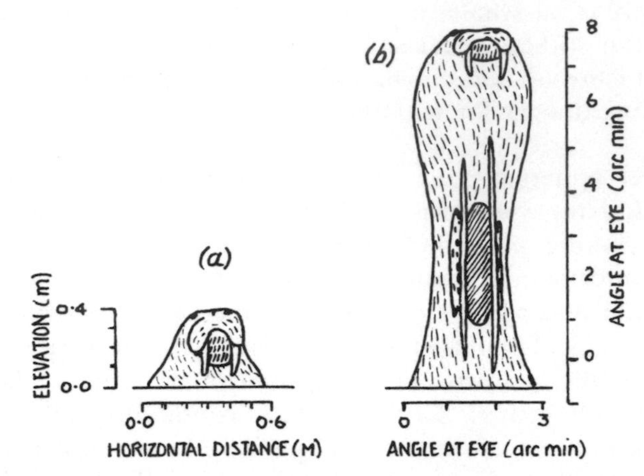

A walrus distorted to look like a merman. *From a drawing by Lehn and Schroeder in* Nature.

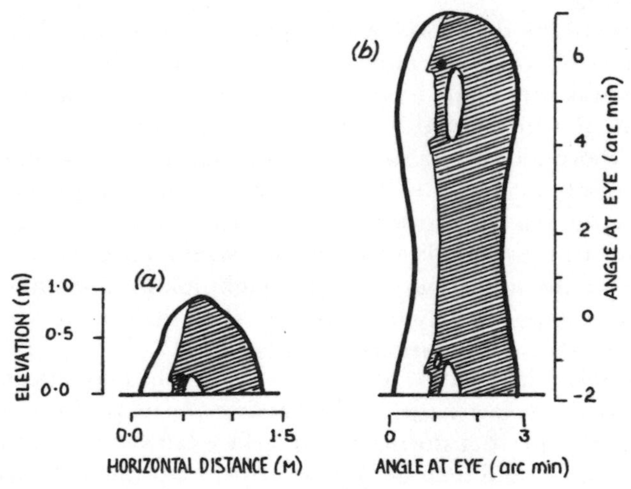

A spy-hopping killer whale distorted to look like a sea monster. *From a drawing by Lehn and Schroeder in the journal* Nature.

these atmospheric effects. The British whaler, William Scoresby, who sailed frequently in Arctic waters, mentions the vertical elongation of the hulls of ships – as if they were heeling over – when certain atmospheric conditions prevailed.

I myself have seen a strange mirage-like atmospheric effect in the Stockholm Archipelago. Islands appear to float in space or have become stretched so that a line of trees appears to be very much taller than it really is.

Lehn and Schroeder's work also casts new light on the way we might interpret other monster reports. The thirteenth-century reports after all, although wrong in their interpretation, are accurate in their description. The *King's Mirror* also described the 'hafgufa' or kraken, a featureless monster of enormous dimensions in the Iceland Sea. It was established in the nineteenth century that the kraken was a giant squid, an animal far from being featureless. Might it be that down through the ages the word 'kraken' has been transferred to another creature and that the original 'kraken' described by medieval sailors was indeed something quite different? The suggestion, which seems quite reasonable, is that the original 'hafgufa' was a submarine volcanic eruption, a natural event likely to be seen in the Iceland Sea and one which would cause considerable disturbance in the water as if some gigantic monster were thrashing about. It would also explain the descriptions, by the likes of Olaus Magnus and Pontippodan in the sixteenth century, of the 'kraken' as big as an island (see page 150). As Lehn and Schroeder wrote in their report in the journal *Nature:*

> In general modern scientists have avoided the study of such anomaly sightings, partially because it is regarded as a non-soluble problem, and partially because it can be so easily distorted.

Let us hope that more scientists, like Lehn and Schroeder, come out of their establishment closets, and intrigue the rest of us with rational and well worked out scientific explanations for some of these mysterious observations.

When we consider the seals and sea lions, the optical distortion phenomenon may be at work here too. Lehn's second computer programme was with a walrus. Again, the 40cm (16in) high walrus head poking above the surface of the water was transformed into a 3m (10ft) high giant with a fierce-looking head on top of a long thin 'neck or waist'.

Under more normal conditions seals can very easily be turned by the untrained eye into sea monsters. The largest living seal, for example, is a monster in itself. It is the male elephant seal *Mirounga sp.*, large specimens of which can weigh up to three tonnes and measure 6.5m (21ft) long.

The largest known elephant seal was one of the southern species *M. leonina* which hauled out to breed each year in Possession Bay, South Georgia; that is until it was shot in 1913. It was 6.9m (22ft 6in) from the tip of its bladder-like nose to the ends of its hind flippers, and was estimated to have weighed 4 – 5 tonnes. On land it could raise its head over three metres above the ground. Most are seen at their breeding beaches on sub-Antarctic Islands and in Tierra del Fuego and southern Argentina.

Occasionally, adult bulls stray to South Africa or New Zealand, often causing considerable consternation when they leave the water and head for the nearest centre of civilization. In 1924, such a bull frightened the heck out of the inhabitants of Smithstown, Cape Province, in South Africa. The local populace responded in the usual way – they shot it. And more recently, in 1987, a large male came ashore on the South Island of New Zealand. It headed straight for a dairy farm and 'fell in love' with the resident cows. After much confusion, it was chased back to sea.

The northern species *M. angustirostris* lives along the Pacific coast of North America and is a likely candidate for some of the 'Caddy' sightings off the coast of British Columbia mentioned in Paul LeBlonde and John Sibert's report (see page 81). Rod Cline's monster in 1948, for example, is possibly a large elephant seal, although as LeBlonde and Sibert point out in their notes, there is no reference to the bulbous nose.

In 1964, Robert Young of Edmonds, Washington, was steaming through Lama Passage, between Cliff Point and Serpent Point on a calm clear morning when his attention was drawn to a log-like object standing upright some 200 – 250 yards away in the water. As he got closer he saw that the sun reflected off dark wet skin, and with binoculars could see that the object was, in fact, the head of a marine animal about the size of a walrus. It had no whiskers or tusks. It submerged vertically. Other people mentioned seeing a similar creature to that seen by Robert Young, but those observations described a 'grippy nose'. Again, that looks a prime candidate for elephant seal.

An incident in October 1969 is also quoted in the LeBlonde-Sibert report. It was sent in by F M Leonard of the tug-boat *Mary Foss* and describes a 3.7 – 4.9m (12 – 16ft) long animal, 1.2m (4ft) wide at the waterline. It had a wrinkled, rubbery nose, about 38-46cm (15 – 18in) wide. The skin was dark brown and the eyes small. The sighting occurred about 3.2km (2mi) north of the Tree Point Lighthouse, 64km (40mi) south of Ketchikan, Alaska, not far from the US-Canada border. It is not usual for elephant seals to be seen in Alaskan waters and the description of the nose gives Mr Leonard's monster away as an adult bull elephant seal.

That anybody should see an elephant seal at all is a miracle in itself, for it was hunted almost to extinction in the 1890s. With protection, however, numbers have increased dramatically. It spends much of its time at sea, only returning to land for breeding during the winter months. And at sea, elephant seals are rarely seen for they spend much of their time diving to great depths.

Recently, researchers from the University of California at Santa Cruz, led by Professor Berney Le Boeuf, attached depth recorders to two adult female elephant seals as they were about to put to sea. When the recorders were recovered the information they contained was quite an eye-opener.

One seal broke the seal deep-diving record by reaching 630m (2,067ft). It turned out that they made more than sixty dives a day, staying down for about twenty minutes at a time, and resting at the surface for only three or four minutes between dives. They also, it seems, continue diving both day and night, and it is calculated that they spend about 90 per cent of their lives under the sea. They sleep, it is thought, by 'cat-napping' as they drift down at an estimated rate of 10 – 20m (33 – 66ft) per minute. The advantage of spending so much time in the deep ocean, apart from giving ample time to gather food, is to avoid predation by surface-living sharks, such as the great white.

Seals are also implicated in sea monster accounts because many commentators speculate that a species of long-necked seal or sea-lion might be living in the sea. It is unlikely, I feel, that there is a new species unknown to science, for all living seals and sea-lions must come on to land or to ice-floes in order to pup. They are, therefore, fairly conspicuous. There is, however, an element of the long-necked seal hypothesis which has some relevance.

Seals are curious creatures in that they can change their shape, particularly the shape of the neck. It is an adaptation to swimming at speed. They have the same number of neck vertebrae as a man or a giraffe, but by stretching the neck out they can almost double its length and make themselves more streamlined. Many a pinniped researcher, as seal expert Sheila Anderson told me, has been caught out by this anatomical contortion. You can pass a seal at what you consider a safe distance only to find a pair of snapping jaws close to your leg.

One seal which not only shows this elongated neck but also has what can only be described as a serpentine head is the leopard seal, *Hydrurga leptonyx.* This cunning and ferocious seal lives in the Southern Ocean where it will patrol a penguin breeding colony waiting to grab the birds when they go fishing or return with their catch. It is an inquisitive animal. Many Antarctic mariners have been surprised when its head pops up beside them and it extends its neck to peer into their boat.

The leopard seal is not as large or bulky as the elephant seal. It has a sleek, steamlined body up to 3m (10ft) long, a large head with powerful jaws and a wide mouth filled with sharp serrated teeth. It is certainly another potential sea monster substitute, but it has size against it. Size, though, is difficult to estimate in a featureless sea where you have no other clues. It is quite possible that many sea monster or sea serpent sightings are inaccurate because of this difficulty. And as soon as the size of many of the creatures featuring in reports down the years comes into question, the selection of quite ordinary identities for supposedly mysterious creatures becomes all the more easy.

A marine mammal that has size on its side, but which is thought to be extinct, is Steller's sea-cow, *Hydrodamalis gigas.* This 9 – 11m (30 – 36ft) long dugong-like creature was thought to have weighed up to 24 tonnes; truly an enormous animal. It lived in the north Pacific, where a population of about two thousand individuals browsed on seaweeds in the shallow waters around the Bering and Copper Islands in the Bering Sea. Although it was known to the inhabitants of the far north, science did not catch up with it until 1741, when Georg Steller described it. Unfortunately, its meat tasted better than that of seals and was preferred by sealers. Within thirty years the entire population was wiped out.

Some marine biologists believe the animal might just have hung on in some remote corners of the Arctic and North Pacific. Soviet whalers, working off Cape Navarin in the north-west Bering Sea, reported seeing what looked like a small group of 6 – 8m long sea-cows in 1962. And a skeleton was found, the first this century, on a Soviet island in 1983. It is clear that such a large and rare creature encountered at sea could easily be logged as an unknown sea monster.

Its smaller cousin, the dugong *Dugong dugong,* which lives in coastal waters from the Persian Gulf to Australia and Papua New Guinea, is the animal that not only gave rise to the myth of the mermaid but also provided us with the identity of the mysterious Ri.

The Ri is an animal described by the Barok people of New Ireland, a province of Papua New Guinea, as having a human-like upper torso and head. The intriguing question infiltrated the cryptozoological community when the Baroks said they recognized dugongs and dolphins and that the Ri was neither of these. During 1979 – 80, Professor Roy Wagner, of the Department of Anthropology, University of West Virginia, discovered references to the Ri when conducting field work. He first heard about the animal from one of the older men of a village and then, at a feast, met a person who saw a Ri every day on the flooded reef at Ramat Bay. It displayed its head and hands to local villagers standing on the shore 'to show that it too was human'.

Wagner was invited to watch for the Ri with his informant. He sat in a tree and waited. He reported what happened next in the Winter 1982 edition of *Cryptozoology*:

> After perhaps a half hour, he became very excited and shouted, 'There it is, there it is!' Gradually, I could discern, several hundred yards out, something large swimming at the surface horizontally. Suddenly, a sawfish jumped immediately in front of it (the range was close enough that I could identify the facial projection), and the dark object submerged and did not reappear.

Wagner questioned more people about the Ri and heard about a village magistrate who came face to face with a Ri on a fishing expedition and threw his spear at it, and a group of boys who found a stranded female Ri, but were so embarrassed at its nakedness that they let it creep back into the sea. A 10-year-old boy described seeing a line of Ri – father, mother and baby – swimming up a stream one moonlit night. An old female Ri was caught near Namatanai and she apparently 'uttered an almost human cry' of pain when she was thrown into the back of a truck and taken off to be killed and eaten – the flesh is considered a delicacy. The Ri's head, so one witness said, was like that of 'a monkey'.

Wagner continued to chronicle the folklore and came to the conclusion that the dugong was not the most obvious animal fitting the description of the Ri. The Baroks, for example, had another name for the dugong – they called them *bo narasi,* meaning 'pig of the sea'. Wagner was perplexed, but as the Ri was not central to his research interests he temporarily dropped the investigation.

In 1983, Wagner returned to New Ireland with a group of cryptozoologists. As the locals had specifically stated that the Ri was not a dugong or any other known marine mammal, the expectation of finding a hitherto unknown species was enough to warrant an investigation. They discovered that in the north of New Ireland the Ri and the dugong are considered to be the same animal. In the south, they were distinct.

The expedition was drawn to the village of Nokon, near Elizabeth Bay on the western side of Cape Natanatamberan where the Ri was known by the name of Ilkai. In Nokon Bay they saw an unidentified animal enter the bay and show its back, though no tail or fins. A little later a mammalian fluked tail was seen, but still no positive identification could be made. A long net was put out in an attempt to trap the creature but all they caught was three reef sharks and numerous other fish.

In writing up their report in *Cryptozoology,* Winter 1983, the team still rejected the dugong, despite spending considerable time comparing the Ri's rolling behaviour at the surface with that of the dugong. The dugong, they felt, was less agile than the Ri. Finless porpoises were considered but eliminated. The mystery continued.

In February 1985, a new, thirteen-strong expedition arrived in New Ireland. They had their own diving boat and were led by Otter and Morning Glory Zell, Tom Williams and Bill Morris, and funded by the Ecosophical Research Association. When they reached Nokon Bay, the Ri or Ilkai put on a good display. It showed its tail flukes and rolled its back. A young Nokon villager was asked about the Ilkai; he pointed to the creature in the bay and said, 'There it is'. A little later the captain of the boat went underwater with snorkel and mask and met the animal. He said it looked like a dugong.

One morning, a dead dugong, killed with a high velocity rifle bullet by a neighbouring villager, was hauled on to the beach. The Ri or Ilkai was, indeed, a dugong. In a report in *The ISC Newsletter,* Tom Williams is quoted as saying:

> The bottom line is that there was a mystery to be solved, like so many others in cryptozoology. We went with the right equipment and solved it. We cryptozoologists solved it ourselves, which is as it should be. It was not solved by the debunkers or skeptics or armchair speculators.

And so, the mystery of the Ri was solved. There are, however, yet more extraordinary tales left to tell. And, if I have carried you along so far without stretching credibility too far, the next couple of examples will certainly have you puzzled.

The first hint that something very strange was going on came in 1883 when the *New Zealand Times* carried an article reporting a 12.2m (40ft) long monster washed up on a Queensland, Australia, shore. Nothing odd about that – humpback whales, which migrate along the outer edge of the Great Barrier Reef, sometimes strand in this area. But what made this carcass unusual is that a 2.4m (8ft) long snout or trunk 'in which the respiratory passages are yet traceable' was found among the remains. No more is known about the body.

Many years later, in 1924, the *Daily Mail* referred to a story published in a provincial South African newspaper in 1922. The eyewitness was Hugh Balance, who had not long previously bought a farm on the Natal coast at Margate. One morning – 1 November, to be precise – Balance was looking out to sea when he saw a commotion in the water. He ran for his binoculars, and, putting them to his eyes, he witnessed two whales fighting with a third unknown creature that raised itself 6m (20ft) out of the water. The gargantuan battle continued, watched by a growing crowd of people on the beach. After about three hours the whales swam away, leaving the third animal floating lifeless on the surface.

This is where belief must be stretched somewhat, for Balance thought the mysterious attacker was none other than a giant polar bear with a long tail. It apparently used its tail to beat off the two assailants. Here, though, we have a fundamental problem, for polar bears do not live in the southern hemisphere, they do not grow to more than 3m (9ft 9in) long (although the largest known polar bear was over 3.4m (11ft) long), and they certainly do not have long tails. The story did not end there.

During the night, the carcass drifted on to the beach near Tragedy Hill and was left there by the receding tide. It was measured and found to be 14.3m (47ft) long, 3.1m (10ft) wide, and 1.5m (5ft) high. At one end was a 3.1m (10ft) long tail and at the other . . . and I pause for breath . . . was a 1.5m (5ft) long and 38cm (15in) diameter trunk. The end of the trunk was shaped like a pig's snout. The body was covered in 46cm (18in) long snow-white hair.

The rotting carcass lay on the beach for about ten days before 32 oxen were harnessed together and attempted to tow it back to the sea. It was left at the water's edge at low tide and was eventually washed out to deeper water. Many local people saw it but, curiously, no scientific report or description has ever been unearthed.

It was not, however, to be the only creature of its type to turn up unexpectedly for, in 1930, the carcass of a smaller 7.6m (25ft) white-furred elephant-like animal with 1m (3ft) long trunk was seen on Glacier Island, Alaska, and in 1944 a 6m (20ft) long *headless* corpse, with the bulk and shape of an elephant, but again covered with white thick fur, was washed ashore at Machrihanish Bay on the west side of Kintyre. The Kintyre carcass was labelled 'polar bear' despite the enormous size, but was it a polar bear? If not, what could it have been?

I don't think we can come to any sensible conclusion about the identity of the various carcasses, but there are a few other factors which might be considered. The Picts – Celtic people who once lived in the northern parts of Britain – were in the habit of carving some of the creatures known to them on stone monuments, and on the Maiden Stone at Meigle in Aberdeenshire there is carved none other than a creature known locally, and misleadingly, as the 'elephant'.

Also, in the recent past, many real living elephants have been found swimming in the sea, many kilometres from land, or floating dead on the surface. There was, for example, an elephant's body washed ashore at Widemouth Bay, not far from Bude in Cornwall, in 1971. Another was found on the beach at Senzu-mura on the Japanese island of Oshima, and two more were washed on to the beach near Wellington in New Zealand in 1955. The Grimsby trawler *Ampulla* cast its nets into the North Sea and came up with – yes, you've guessed it – an elephant! And in 1982 the crew of an Aberdeen fishing boat were surprised to meet a dead elephant about 50km (32mi) offshore, also in the North Sea.

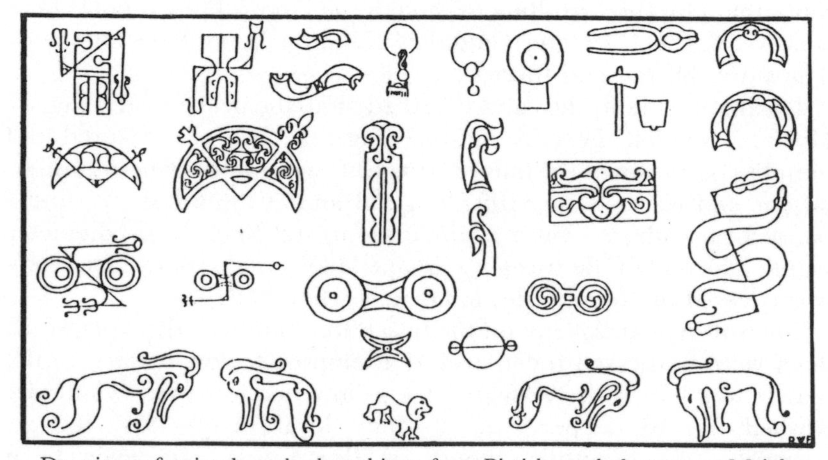

Drawings of animals and other objects from Pictish symbol stones at Meigle, Aberdeenshire including, on the bottom line, the 'swimming elephants'.

In the *New Scientist* of 2 August 1979, there was even a photograph taken by Admiral R Kadirgama of an elephant swimming off the coast of Sri Lanka. Many cryptozoologists and sceptics alike have pointed out a striking resemblance between the swimming elephant, with its trunk waving in the air, and reports of sea serpents and other sea monsters.

Could it really be that the numerous sightings of sea monsters are simply wayward elephants? The Mary F photographs, for example, published in the *Falmouth Packet,* look a little like an elephant's trunk waving in front of an elephant's head. And, don't forget, a real elephant came ashore not far away, at Bude.

Also, if we question some of the features of the Margate Monster we may come a little closer to understanding the events of the time.

Might the skin, tough like that of the basking shark, be rubbed off, exposing the underlying muscle fibre which becomes teased out, again like the shark, to produce 'fur'? Would not the ears be the first external structures to be worn away? Would the tail, with its core of tail vertebrae and the trunk with its tough muscles resist erosion? Might the body be swollen when waterlogged so as to give an exaggerated size? And who did the measuring anyway? Is there no reference available? Might it be that the creature was not alive in the water, but had already been dead for some time? Was the carcass, then, an easy meal for a pair of killer whales which, competing with a group of sharks, caused considerable commotion in the water? The questions are reasonable, but, I am afraid, they have no answers.

As for the Pictish 'swimming elephant', it looks remarkably like a dolphin, with a dolphin's snout, flippers, and a curly 'blow' emanating from the top of its head.

So much, then, for elephants, and so much for fish, reptiles and other mammals as likely 'known' identities for sea serpents and other mysterious marine monsters. But there is one area left to consider. Is it possible that an ancient prehistoric animal, known only from the fossil record, could have survived all these millions of years and is living today in the remote parts of the world's oceans? There was one occasion, recently, when we were forced to consider this very possibility.

In June 1983, British schoolboy Owen Burnham was on holiday at a Gambian seaside resort, known as Bungalow Beach, when he discovered a carcass washed up by the tide. He wrote to the readers' letters page of *BBC Wildlife Magazine* where his description of the beast was published:

> It was about 15ft long and had a dolphin-like head with long jaws and 80 teeth. Its body was large, with no blowhole and no distinct neck, and there were two flippers behind its head and two in the pelvic region (the latter were damaged and one was torn off). The animal had a five-foot long pointed tail and altogether resembled a very small Kronosaurus (an Australian plesiosaur – 'near lizard – of between 136 and 65 million years ago). An expert at the Natural History Museum suggested that it might be a dolphin whose tail flukes had worn off, but there were no signs of damage and I am not convinced by this theory.
>
> Can anyone help identify the animal?

Burnham also enclosed some sketches of the beast. The magazine approached marine experts at Cambridge University but they were unable to draw any firm conclusions. They suggested that, if the description was an accurate one, it might indicate a 'living fossil' of something akin to a mososaur, plesiosaur or archaeocete. The magazine's readers were then asked to contribute their ideas as to what the creature might be, and a couple of months later I was asked to pull together their replies and sum up the findings.

Further investigation revealed a little more information about the 'beaked beast of Bungalow Beach', as it became known. The body was about 1.5m (5ft) wide, with a smooth scaleless skin, brown on top and white below. The front 76cm (30in) of the 127cm (50in) long dolphin-like head was drawn out into long jaws. On the upper surface of the snout there were two small 'nostrils'. The head was slightly dome-shaped. There was no dorsal fin. The front paddle-like flippers were about 46cm (18in) long. The body contained blubber.

Unfortunately, Burnham did not have a camera with him and failed to take any samples for later investigation. In fact, local fishermen hacked off the creature's head and took it away to sell to tourists as a 'crocodile' skull. So, we are left with the descriptions and the drawing. If they are accurate, then we have a biological problem, for no creature with these characteristics is known to be living in the sea today.

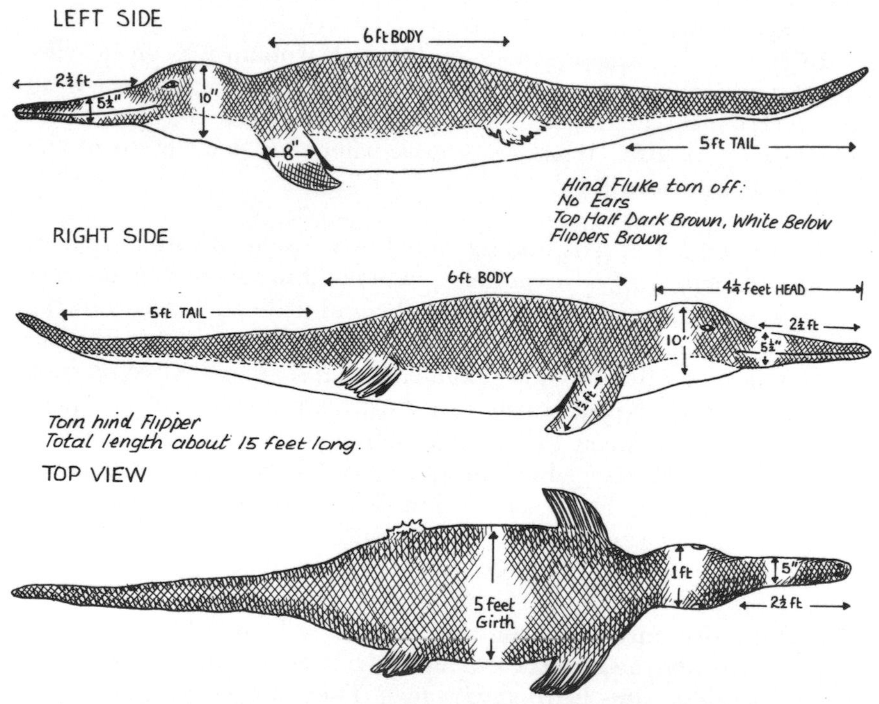

The 'beaked beast of Bungalow Beach' in the Gambia. *From drawings by Owen Burnham.*

There is, however, one animal that *almost* fits the description – Shepherd's beaked whale *Tasmacettus shepherdi.* This most primitive of the beaked whales is known from about a dozen stranded individuals, all found in the southern hemisphere. The suggestion came from Dr Andrew Milner, of the University of London, a specialist in fossil amphibians and reptiles. He pointed out that although the rest of the beaked whales have few teeth, Shepherd's whale has a set of about eighty.

Another suggestion was that the carcass was that of a primitive whale or archaeocete. These ancient creatures also had lots of teeth, but they differed from the Gambian carcass in not having the slightly 'domed' forehead that Burnham described. This would indicate the presence of a 'melon' (used in echo-location to focus sound), and is a characteristic of some modern whales and dolphins.

The beaked whales have small pectoral flippers and long, slender tails with flukes. If the carcass's flukes had been scuffed or bitten off, there would be little visible damage. Its single, damaged rear flipper could have been a flap of torn skin. A small dorsal fin and a narrow crescent-shaped blowhole, resembling a tuck of skin, could easily have been missed. If, however, Burnham's description was accurate, then we must turn to more fanciful suggestions.

Mososaurs and pliosaurs were short-necked and long-bodied marine reptiles that hunted the seas in Cretaceous times but, like their contemporaries the dinosaurs, became extinct about 65 million years ago. Mososaurs had large fang-like teeth and were thought to have had scales on their skin, much like today's monitor lizards, which are considered to be their living relatives. The pliosaurs, though, had an elongated head, many small teeth and smooth skin.

Another contender, proposed by Karl Shuker, was a group of primitive marine crocodiles known as thalattosuchians, which disappeared from the fossil record in the late Jurassic – Lower Cretaceous. They had long jaws and many teeth, a smooth skin, and 'paddles' instead of crocodile-like claws. But their long tail ended in a downward-pointing section with a rayed-fin on the upper side. Reconstructions, based on their fossil remains look remarkably like Burnham's drawings (if you imagine the fragile tail fin scuffed away).

In the not too distant past, there have been other crocodile-like or primitive whale-like sightings, described in Heuvelmans' *In the Wake of Sea Serpents,* which may hint at ancient reptiles surviving to the present day. In 1877, for example, the US ship *Sacramento* was in mid-Atlantic at latitude 31° 50'N and longitude 37°W when it encountered a 18.3m (60ft) long snake-like creature with an alligator's head. The head, so the observer said, was raised out of

Primitive crocodiles in a Jurassic sea. *From drawings in* Scientific American.

the water by about 1m (3ft) – a feature not usually observed in salt-water crocodiles, the only known explanation for such a sighting at sea. Salties lie low in the water except when gulping down particularly large prey.

A similar sighting was referred to in W H Marshall's *Four Years in Burma.* He describes a 13.7m (45ft) long crocodile swimming rapidly past the ship *Nemesis* heading for Rangoon. Large salt-water crocodiles are not uncommon in those parts and so, no doubt, it was shrugged off as a large specimen. Marshall, however, adds one more observation which, on reflection, is odd. He said the head and half the body were out of the water – unlikely behaviour for a salt-water crocodile. A thalattosuchian, however, with the downward end to its tail could, perhaps, push the front part of its body clear of the water.

The *Ambon* sea serpent of 1904. *From a drawing by J Vollewens.*

On the other hand, the sighting from the *Ambon* on passage from the Gulf of Aden to the Red Sea and recorded by the third officer – a strange animal that raised its head above the water, looked like a cayman, was quite smooth above and below, had long jaws with pointed teeth top and bottom, small eyes, and a fin on the back – was probably a Shepherd's beaked whale. The sighting was in 1904 when the whale had not been recognized by science.

So, was the Gambian carcass that of a saurian reptile, of a kind thought to have become extinct many millions of years ago? The odds are against it, but if it was a 'living fossil', it would not be the very first time that such a discovery had been made. As I mentioned at the outset, the coelocanth, megamouth and a host of other unusual marine animals have presented themselves unexpectedly to be recognized by modern science. Indeed, as I write, another giant leatherback turtle – possibly the largest ever found – has been washed ashore on the Welsh coast and will be on display in the National Museum of Wales in Cardiff, and a third 4m (13ft) long specimen of megamouth has been found alive on a beach in Western Australia. It came ashore near the resort town of Mandurah, and local residents tried in vain to push it back out to sea; it became beached again and died. I wonder how many more of these extraordinary marine creatures we can expect to find.

There is, however, one more shark which I would like to mention. It is, perhaps, the most exciting and the most spine-chilling of them all.

In the rocks laid down in the middle and late Tertiary times palaeontologists found large sharks' teeth. They were as much as 15cm (6in) in length with finely serrated edges, and resembled the teeth of the great white shark. They belonged to a shark known scientifically as *Carcharodon megalodon,* an animal that reached its greatest size during the Miocene period; it was a shark that was truly a giant among sharks.

In 1982, John Maisey was asked to reconstruct the jaws of this enormous creature for the Smithsonian Institution in Washington DC. An amateur fossil collector, Peter Harmatuk, found a partial set of megalodon teeth in Miocene rocks in a North Carolina phosphate quarry. By using this data and the measurements we have today of the teeth and jaws of great white sharks, Harmatuk was able to draw up a blue-print for the ancient shark. He worked out that it was gigantic. It had a gape of about two metres (6½ft), and was up to 30m (98ft) in total body length. It was the largest and most powerful predatory fish that ever lived.

Imagine, then, the shock when scientists dredging the bottom of the Pacific Ocean, earlier this century, discovered two 10cm (4in) long megalodon teeth that were what can only be described as geologically 'fresh'. One was estimated to be 24,000 years old –

roughly the time of the Lascaux cave paintings. The other was just 11,000 years old and therefore belonged to a gigantic shark that swam in the Pacific Ocean at the same time as man was migrating from Asia into North America. Could this enormous predator still be lurking in the ocean depths?

As my grandfather used to say as we scanned the empty shore on the South Devon coast, 'We shall have to wait and see what the tide brings in'.

Bibliography

For the student who wishes to pursue further an interest in marine cryptozoology there are several books which are essential reading; *In the Wake of Sea Serpents* by Bernard Heuvelmans (London, Rupert Hart-Davis, 1968)

The Case for the Sea Serpent by Rupert Gould (London, Philip Allan, 1930)
The Great Sea-Serpent by Antoon Cornelis Oudemans (London, Luzac & Co., 1892)
The Lungfish, the Dodd, and the Unicorn by Wiley Ley (New York, Viking, 1948)

For an introduction to cryptozoology, I suggest:
Searching for Hidden Animals by Roy Mackal (New York, Doubleday, 1980)

For the identification of marine creatures:
The Seafarer's Guide to Marine Life by Paul Horsman (London, Croom Helm, 1985)
A Field Guide to the Mediterranean Sea Shore by Wolfgang Luther and Kurt Fielder (London, Collins, 1976)
Collins Guide to the Sea Fishes of Britain and North-Western Europe by Bent Muus and Preben Dahlstrom (London, Collins, 1974)
Whales of the World by Lyall Watson (London, Hutchinson, 1981)

For marine life and other animal superlatives:
The Guinness Book of Animal Facts and Feats by Gerald Wood (London, Guinness Superlatives, 1982)

For up-to-date information on unusual marine creatures and other wildlife news I suggest: subscribing to *BBC Wildlife Magazine*, PO Box 168, Tunbridge Wells, Kent TN2 3UX; and, joining the International Society for Cryptozoology, ISC Secretariat for Europe, 25 chemin de Trembley, 1197 Prangins, Switzerland.

Index

Tenth Edition

KINESIOLOGY

Scientific Basis of Human Motion

Nancy Hamilton, Ph.D.
Associate Professor, University of Northern Iowa

Kathryn Luttgens, Ph.D.
Professor Emerita, Northeastern University

Boston Burr Ridge, IL Dubuque, IA Madison, WI New York San Francisco St. Louis
Bangkok Bogotá Caracas Kuala Lumpur Lisbon London Madrid Mexico City
Milan Montreal New Delhi Santiago Seoul Singapore Sydney Taipei Toronto

McGraw-Hill Higher Education

A Division of The **McGraw-Hill** *Companies*

KINESIOLOGY: SCIENTIFIC BASIS OF HUMAN MOTION
TENTH EDITION

Published by McGraw-Hill, a business unit of The McGraw-Hill Companies, Inc., 1221 Avenue of the Americas, New York, NY 10020.

This book is printed on acid-free paper.

1 2 3 4 5 6 7 8 9 0 DOC/DOC 0 9 8 7 6 5 4 3 2 1

ISBN 0-07-232919-X
ISBN 0-07-112243-5 (ISE)

Vice president and editor-in-chief: *Thalia Dorwick*
Executive editor: *Vicki Malinee*
Senior developmental editor: *Michelle Turenne*
Senior marketing manager: *Pamela S. Cooper*
Project manager: *Mary Lee Harms*
Production supervisor: *Enboge Chong*
Coordinator of freelance design: *Rick D. Noel*
Cover designer: *Jamie A. O'Neal*
Cover image: © *Imagedrome, Inc.*
Senior photo research coordinator: *Lori Hancock*
Senior supplement producer: *David A. Welsh*
Media technology producer: *Judi David*
Compositor: *Shepherd, Inc.*
Typeface: *10/12 Times Roman*
Printer: *R. R. Donnelley & Sons Company/Crawfordsville, IN*

Library of Congress Cataloging-in-Publication Data

Hamilton, Nancy (Nancy Patricia), 1946–   .
Kinesiology : scientific basis of human motion / Nancy Hamilton, Kathryn Luttgens. — 10th ed.
p. cm.
On the previous ed. Luttgens appears first.
Includes bibliographical references and index.
ISBN 0-07-232919-X — ISBN 0-07-112243-5
1. Kinesiology. I. Luttgens, Kathryn, 1926–   . II. Title.

QP303 .L87 2002
612.7'6—dc21

2001030368
CIP

INTERNATIONAL EDITION ISBN 0-07-112243-5

The Internet addresses listed in the text were accurate at the time of publication. The inclusion of a website does not indicate an endorsement by the authors or McGraw-Hill, and McGraw-Hill does not guarantee the accuracy of the information presented at these sites.

www.mhhe.com